Lecture Notes in Artificial Intellige

Subseries of Lecture Notes in Computer Science
Edited by J. G. Carbonell and J. Siekmann

Lecture Notes in Computer Science
Edited by G. Goos, J. Hartmanis and J. van Leeuwen

Springer
Berlin
Heidelberg
New York
Barcelona
Hong Kong
London
Milan
Paris
Tokyo

Andrea Omicini Paolo Petta
Robert Tolksdorf (Eds.)

Engineering Societies in the Agents World II

Second International Workshop, ESAW 2001
Prague, Czech Republic, July 7, 2001
Revised Papers

Series Editors

Jaime G. Carbonell,Carnegie Mellon University, Pittsburgh, PA, USA
Jörg Siekmann, University of Saarland, Saarbrücken, Germany

Volume Editors

Andrea Omicini
DEIS
Università di Bologna, Alma Mater Studiorum
Facoltà di Ingegneria della Romagna, Sede di Cesena
Via Rasi e Spinelli 176, 47023 Cesena (FC), Italy
E-mail: aomicini@deis.unibo.it

Paolo Petta
Austrian Research Institute for Artifiical Intelligence
Software Agents and New Media Group
Schottengasse 3, 1010 Vienna, Austria
E-mail: paolo@ai.univie.ac.at

Robert Tolksdorf
Technische Universität Berlin
Fachbereich 13 - Informatik - FR 6-10
Franklinstr. 28/29, 10587 Berlin, Germany
E-mail: tolk@cs.tu-berlin.de

Cataloging-in-Publication Data applied for

Die Deutsche Bibliothek - CIP-Einheitsaufnahme

Engineering societies in the agents world II : second international workshop ; revised papers / ESAW 2001, Prague, Czech Republic, July 7, 2001. Andrea Omicini ... (ed.). - Berlin ; Heidelberg ; New York ; Barcelona ; Hong Kong ; London ; Milan ; Paris ; Tokyo : Springer, 2001
(Lecture notes in computer science ; Vol. 2203 : Lecture notes in artificial intelligence)
ISBN 3-540-43091-1

CR Subject Classification (1998): I.2.11, I.2, C.2.4, D.1.3, D.2.2, D.2.7, D.2.11

ISSN 0302-9743
ISBN 3-540-43091-1 Springer-Verlag Berlin Heidelberg New York

Springer-Verlag Berlin Heidelberg New York
a member of BertelsmannSpringer Science+Business Media GmbH

http://www.springer.de

Typesetting: Camera-ready by author, data conversion by Christian Grosche, Hamburg
Printed on acid-free paper SPIN 10840729 06/3142 5 4 3 2 1 0

Preface

The idea to initiate a series of workshops to be entitled "Engineering Societies in the Agent's World" (ESAW) originated in late 1999, among members of the working group on Communication, Coordination, and Collaboration of the Intelligent Information Agents special interest group of AgentLink, the European Network of Excellence for Agent-Based Computing. By that time, the convergence of scientific and technological progress in numerous areas, including software engineering, distributed problem solving, knowledge-based systems, and dynamic pervasive networking had gained significant momentum. As a result, the reality of multiagent systems was now a given.

In the eyes of the proposers and supporters of ESAW, these developments led to a new and manifest need, that was being left all but uncovered by the existing range of conferences and meetings. A platform that would overcome the disparate roots of the multiagent systems field by placing a clear focus on an integrative level of analysis, namely: artificial societies populating a world encompassing the natural and the artificial, comprising autonomous entities and their environment.

The chosen point of view identifies the topics of autonomy, dynamics, interaction patterns, agent lifeworlds, and coordination as central notions. Reflections of these are sought in the "traditional" multiagent systems areas of theoretical foundations for conceptualization, abstractions and models for design, methods and tools for development, supporting hardware technologies for realization, and procedures for deployment and governance. For this purpose, contributions are gathered from an open range of relevant areas in the humanities, natural sciences and engineering, technology and industry, extending those already firmly established within the multiagent systems community. Together with application experiences already gathered in building the first agent societies, these shall enable and facilitate pertinent tasks, including

- development of coordination models and technologies for engineering of agent societies
- engineering of social intelligence and emergent behaviors in MAS
- design of enabling infrastructures for agent societies
- implementation of centralized and decentralized forms of social control
- impacts of visibility and individuality of agents
- verification, validation, and certification of agent societies
- design of interaction/coordination patterns for agent societies
- solving issues of mobility, security, authority, and trust in agent societies
- development of methods, tools, and artifacts for engineering agent societies

ESAW thus commits to the use of the notion of multiagent systems as seed for animated constructive discussions of high quality about technologies, methodologies, and models for the engineering of complex distributed applications. Concentrating on social aspects of multiagent systems, an emphasis is placed on technology and methodology issues, while also welcoming theoretical and empirical contributions with well-documented connections to these core subjects. The series enjoyed a successful start as a workshop at ECAI 2000 in Berlin, with the results documented in the predecessor of the present volume.[1]

The second edition of this workshop series, ESAW 2001 was held at the premises of the Czech Technical University in Prague (CTU) on July 7, 2001 in co-location with the second AgentLink II general meeting and the MASAP 2001 courses on Multi-Agent Systems and Applications: the 3rd European Agent Systems Summer School (EASSS 2001), and ECCAI's Advanced Course on Artificial Intelligence (ACAI-01). This offered young researchers the opportunity to appreciate first-hand the current consensus as well as topics of debate among the over 50 attendants participating in the discussions until the late afternoon. This volume covers extended versions of papers presented at the workshop, which include results of discussions and insights gained from other presentations. The content covers the areas of Foundations of Engineering with Agents, Logic and Languages for MAS Engineering, Agent Middleware, and Applications.

Three main issues emerged from the presentations: *agent languages* – here the rather well-known debate about homogeneity vs. heterogeneity ranged from direct utilization of natural language technology over results from ontological engineering such as in connection with the Semantic Web initiative, existing standards such as FIPA-ACL, all the way to novel proposals such as Entish language for agent "souls"; *context* – aspects of the environment making up the agents' lifeworld, context of interaction, available and shared resources, linguistic context; and *responsibility* – in societies of differing degree of openness. These can be captured in the following three questions: What characteristics distinguish a language that is adequate for agents? What does the notion of "context" have to cover for agents and multiagent systems? Who is responsible for the behavior and actions of agents and multiagent systems?

These questions were elaborated and supplemented by contributions presented in an animated plenary discussion that wrapped up this year's edition of the ESAW workshop. There, the central ideas of ESAW as presented above were confronted once more. Under the heading *"The many ends in the Agents' World to make meet: which, whether, how?"*, a probing was undertaken for common ground, but also prominent obstacles and clear divides, based upon the problems and successes, fears and hopes, with respect to one's *own* experience and the *other's* domains. In this context, it was initially made clear that "us" and "them" were intended to be represented by top-down MAS designers progressing from single multi-agent systems to the new challenges posed by "multi-MAS"; bottom-up Distributed AI and MAS designers identifying mappings from traditional system structures and properties to situated distributed solutions; middleware designers investigating how to provide services of increasing complexity; researchers in organization theory and game theory, identifying and characterizing abstract notions and measures to structure and analyse the new domain of the societies in the agents' world; researches in the social sciences and coordination theory, proposing requirements and models in terms of mechanisms and infrastructural constituents; application designers wondering how to predictably bring about all the many "ilities" their case at hand imposes; industrials concerned about how to ensure steady and predictable progress so as

[1] Omicini, A., Tolksdorf, R., and Zambonelli, F. (Eds.): Engineering Societies in the Agents World, First International Workshop, ESAW 2000, Berlin, Germany, August 21, 2000. Revised Papers. LNAI 1972, Springer-Verlag, 2000.

to turn technological promises into a hard reality; and finally the individual perspective of each workshop attendant. Among the many issues brought forward (and to be covered in due detail elsewhere), we summarize a few representative ones in the following.

A basic issue is posed by the need for a clear specification of the notion of "society" in the agents' world, similar to the request for a definition of "context" spelled out before. While it should capitalize upon what is known about "real" societies, the characterization of an agent society should not be constrained by it, given that artificial societies and their lifeworlds do not only have to be modeled but also constructed from scratch. In particular, giving in to the temptation of anthropomorphism may effectively prevent the discovery of new, more pertinent and effective approaches. As a consequence, the scientific issue of modeling and the engineering issue of construction at least for the time being cannot be clearly separated from each other. Relatedly, real world "Embedded Internet" applications have to deal with potentially hostile and changing environments. These aspects have arguably so far been insufficiently addressed in MAS research and development. A shift of focus would be required therefore, from problem solving to providing societal support that takes the changing and potentially hostile environmental context into account. Along with consideration of the characteristics of individual agents and available means for coordination, this should lead to offering an adequate range of both offensive and defensive operations to different types of threats, including the issues of disclosure and integrity. This call to move "good societies" stands in interesting contrast to other observations offered in the literature suggesting that the danger posed by misbehaving entities in a MAS might be a non-issue after all.

As another significant entry, the importance of proper consideration of business modeling was put forward. The value of MAS as a viable alternative to the breaking down of chain-based services being experienced today has to be demonstrated. This must be accompanied with the provision of adequate tools crucial for business takeup. Also in this context, the co-evolution of technologies; laws, norms, and policies regulating (human) societies and organizations; and habits and trusts in individuals must not be disregarded. The social perspective on MAS also has to stand up to comparisons to other approaches and prove its validity. This asks for the identification of proper ways to characterize performances and system behaviors and procedures for the measurement and comparison of the qualities of different solutions. Also in this regard, the social view on multiagent systems is mostly still waiting for its acceptance as a first-class entity and element of analysis. In fact, the very relationship of the notions of artificial agent and artificial society is still awaiting a detailed analysis. It remains to be seen to what extent we may have already embarked upon a transition from the agent-based to a next, higher, self-contained paradigm.

We hope that these few words managed to convey a little bit of the enthusiasm and liveliness of debate that characterized the second edition of ESAW, confirming and even surpassing the success of its predecessor. Its results, compiled in the present volume, form an important step towards the establishment of a rampant new young shoot in multi-agent systems research. We look forward with excitement to the results to be presented at ESAW 2002.

In closing, we would like to acknowledge the exemplary local support provided by Hana Krautwurmova, Olga Stepankova, and Vladimir Marik, and the kind sponsorship of the event offered by the Austrian Society for Artificial Intelligence (ÖGAI). We also thank the members of the program committee for insuring the quality of the workshop program by kindly offering their time and expertise so that each contribution could undergo triple reviewing.

October 2001

Andrea Omicini
Paolo Petta
Robert Tolksdorf

Organization

Organizers and Chairs

Andrea Omicini	*Università di Bologna, Italy*
Paolo Petta	*Austrian Research Institute for AI, Austria*
Robert Tolksdorf	*Technische Universität Berlin, Germany*

Program Committee

Cristiano Castelfranchi (Italy)
Paolo Ciancarini (Italy)
Helder Coelho (Portugal)
Yves Demazeau (France)
Rino Falcone (Italy)
Tim Finin (USA)
Rune Gustavsson (Sweden)
Chibab Hanachi (France)
Matthias Klusch (Germany)
Yannis Labrou (USA)
Jiri Lazansky (Czech Republic)
Lyndon C. Lee (UK)
Andrea Omicini (Italy)
H. Van Dyke Parunak (USA)
Michal Pechoucek (Czech Republic)
Paolo Petta (Austria)
Jeremy Pitt (UK)
Agostino Poggi (Italy)
Antony Rowstron (UK)
Luciano Serafini (Italy)
Christophe Sibertin-Blanc (France)
Munindar P. Singh (USA)
Paul Tarau (USA)
Robert Tolksdorf (Germany)
José M. Vidal (USA)
Franco Zambonelli (Italy)

Table of Contents

Foundations of Engineering with Agents

Logics and Languages for MAS Engineering

Agent Middleware and Applications

Categories of Artificial Societies

Paul Davidsson

Department of Software Engineering and Computer Science
Blekinge Institute of Technology, 372 25 Ronneby, Sweden
paul.davidsson@bth.se

Abstract. We investigate the concept of artificial societies and identify a number of separate classes of such societies. These are compared in terms of openness, flexibility, stability, and trustfulness. The two most obvious types of artificial societies are the open societies, where there are no restrictions for joining the society, and the closed societies, where it is impossible for an "external agent" to join the society. We argue that whereas open societies supports openness and flexibility, closed societies support stability and trustfulness. In many situations, however, there is a need for societies that support all these aspects, e.g., in systems characterized as information ecosystems. We therefore suggest two classes of societies that better balance the trade-off between these aspects. The first class is the semi-open societies, where any agent can join the society given that it follows some well-specified restrictions (or, at least, promises to do so), and second is the semi-closed societies, where anyone may have an agent but where the agents are of a predefined type.

1 Introduction

A collection of software entities interacting with each other for some purpose, possibly in accordance with common norms and rules, may be regarded as an *artificial society*. This use of the term "society" is analogous to human and ecological societies. The role of a society is to allow the members of the society to coexist in a shared environment and pursue their respective goals in the presence of others. As a software entity typically acts on the behalf of a person or an institution, i.e., its *owner*, we will here refer to these entities as *agents*. This use of the term "agent" is somewhat more general than is common. However, since the principles we will discuss are general, covering all kinds of (semi-)autonomous software processes, there is no reason for limiting the discussion to "proper" software agents.

There are a number of other notions used to refer to organizational structures of software agents, e.g*groups* , *teams*, *coalitions*, and *institutions* (cf. for instance [2, 4, 9]). Since a society may contain any number of institutions, coalitions, teams, groups, and individual agents, the concept of society belongs to a higher organizational level than these structures. Also, whereas a society is neutral with respect to co-operation and competition, coalitions are formed with the explicit intention of co-operation. Similarly, a team is a group in which the agents have a common goal. The difference

A. Omicini, P. Petta, and R. Tolksdorf (Eds.): ESAW 2001, LNAI 2203, pp. 1-9, 2001.

between a group or a team and an institution is that an institution has a legal standing distinct from that of individual agents.

Artikis and Pitt [1] have provided a formal characterization of an agent society that includes the following entities:

- a set of agents,
- a set of constraints on the society,
- a communication language,
- a set of roles that the agents can play,
- a set of states of affairs that hold at each time at the society, and
- a set of owners (of the agents).

They describe the set of constraints as "constraints on the agent communication, on the agent behaviour that results from the social roles they occupy, and on the agent behaviour in general." Another way describing the set of constraints is that they constitute the norms and rules that the agents in the society are supposed to abide. When appropriate, we will refer to the above list of entities when discussing different types of societies. In addition, we will here regard yet another entity, namely the *owner of the society*. By this we mean, the person or organization that have the power to decide which agents may enter the society, which roles they are allowed to occupy, what communication language should be used, and the set of constraints on the society. Note, however, that all societies may not have an owner and that there is a difference between the owner and the *designer* of an agent or a society [6].

Depending on its purpose, an artificial society needs to support the following properties to different degrees:

- *openness*, i.e., the possibilities for agents to join the society
- *flexibility*, i.e., the degree to which agents are restricted in their behavior by the society
- *stability*, i.e., predictability of the consequences of actions, and
- *trustfulness*, i.e., the extent to which the owners of the agents trust the society (which may be based on e.g. mechanisms for enforcing ethical behavior between agents).

These properties are not completely independent, e.g., the trustfulness of a society is typically dependent of its stability.

An emerging view of future distributed software systems is that of *information ecosystems*. These are populated by *infohabitants*, i.e., (semi-)autonomous software entities typically acting on the behalf of humans. According this vision, which is inspired by biotic ecosystems, the information ecosystem should be able to adapt to changing conditions, easily scale up or down, and have an openness and universality. (This vision is shared by the Future and Emerging Technologies initiative "Universal Information Ecosystem" within the Information Society Technologies program of the European Commission (see http://www.cordis.lu/ist/fetuie.htm).) Thus, information ecosystems correspond to artificial societies with a high degree of openness, flexibility, stability, and trustfulness.

In this paper, which builds upon earlier discussions [3], we will categorize artificial societies based on their degree of openness. We start by describing two basic types of artificial societies, the open and closed societies, and discuss their strengths and weaknesses. Based on this discussion we suggest two other types of artificial societies, semi-open and semi-closed, which are able to balance the strengths and weaknesses of open and closed societies. Finally, we discuss the trade-off between openness, flexibility, stability and trustfulness.

2 Open and Closed Agent Societies

Two basic types of artificial societies are *open societies,* where there are no restrictions for joining the society, and *closed societies,* where it is impossible for an "external agent" to join the society.

2.1 Open Societies

In principle, it is possible for anyone to contribute one or more agents to an open society without restrictions. An agent joins the society simply by starting to interact with some of the agents of the society.

If we characterize an open society with respect to the four desired properties listed in the introduction, it supports openness and flexibility very well, but it is very difficult to make such a society stable and trustful. For instance, it is not possible to control the set of constraints or monitor whether the agents abide these. In fact, it is not possible to determine the set of agents in any effective way. Within an open society, the only structure is typically just a generally accepted communication language and a limited set of roles. Of course, it is possible to consider artificial societies that do not even have this, which we may call *anarchic societies*.

The most obvious example of an open artificial society is the World Wide Web (WWW), where the set of members of the society consists of the set of WWW-browser processes together with the set of WWW-server processes that are connected to the Internet. HTTP (Hypertext Transfer Protocol) is the communication language. The number of roles is limited to clients, i.e., the browsers, and servers. Finally, the set of owners is either the owners of the machines on which the browser and server processes are run, or the persons/institutions that started the browser and server processes. The openness of the society is obvious in this case; anyone with an Internet connection is allowed to start a browser process or a server process and join the artificial society defined by the WWW without any restrictions.

2.2 Closed Societies

Closed agent societies are typically those where a Multi-Agent System (MAS) approach is adopted by a team of software developers to implement a complex software system. The MAS is designed to solve a set of problems, specified by the society owner. The solving of these problems is then distributed between the agents of the MAS. It is not possible for an "external agent" to join the society. Zambonelli et al. [10] refer to this type of systems as "distributed problem solving systems" and describe them as "systems in which the component agents are explicitly designed to co-operatively achieve a given goal." They argue that the set of agents is known a priori, and all agents are supposed to be benevolent to each other and, therefore, they can trust one another during interactions. In open systems, on the other hand, agents are "not necessarily co-designed to share a common goal" and cannot be assumed to be benevolent.

The concept of closed agent society corresponds to the large majority of existing MAS. An advantage of closed societies is that it is possible to precisely engineer the society, e.g., specify exactly which agents interact, and why etc. Consequently, closed societies provide strong support for stability and trustfulness. However, they are able to provide very little openness and flexibility. Just as we identified anarchic societies as extreme type of open society, it is possible to identify and extreme type of closed societies, namely, the *fixed societies*. In a fixed society, all agents are created in the initialization phase, whereas in a close society, the owner/designer of the society may create new members of the society during execution time.

Our conclusion is that neither open nor closed societies to a sufficient degree support all of the desired features, i.e., flexibility, openness, stability, and trustfulness, that are required for the implementation of large class of software systems, e.g., information ecosystems. We will now investigate two types of artificial societies that are more suitable for this type of systems, namely "semi-open" and "semi-closed" societies to which anybody may contribute an agent, but where entrance to the society is restricted and behavior may be monitored by some kind of institution:

- in *semi-open* societies, the agents may be implemented and run locally
- in *semi-closed* societies, the agents are implemented and run on remote servers.

3 Semi-open Artificial Societies

In what we will call semi-open artificial societies, there is an institution that function as a gate-keeper. Agents wanting to join the society contacts the institution to whom it promises to follow the set of constraints of the society. The institution then makes an assessment whether the agent is trustworthy and eligible and decides whether to let it join the society or not. (See Figure 1.) It is, of course, possible to differentiate between classes of trustworthiness so that agents that are considered more trustworthy are given access to more services etc than agents considered less trustworthy.

A type of institution that implements this functionality to some extent is the *portal* concept as used in SOLACE [5]. In addition to keeping track of the entities and services of the society, it can be used to "ensure that requirements such as security, integrity, and privacy of information related to the services associated with a particular portal is effectively taken care of."

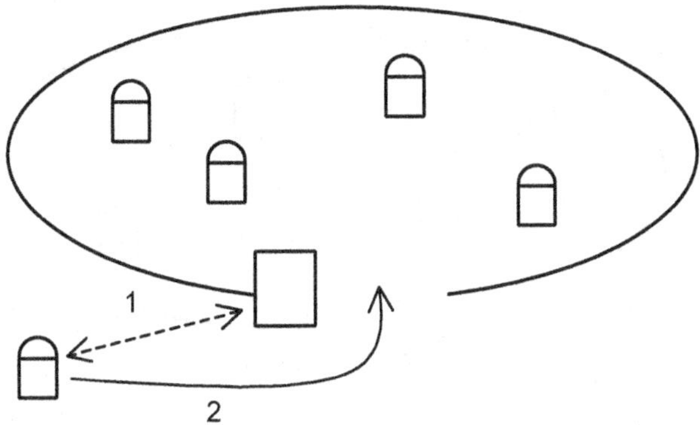

Fig. 1: Logical view of an agent entering a semi-open society. Agents may be arbitrarily distributed in the physical space.

In fact, there are already a number of distributed information systems that resemble semi-open societies. For example, consider peer-to-peer systems [8], such as the Internet-based Napster service (cf. www.napster.com) which let users share their achieves of mp3 files containing music. Each user must use a particular type of Napster software. In order to get access to other users' files, a Napster software process needs to connect to a central server, which then may let the process join the society. If the process succeed to join the society, it will be able to interact with other users' Napster software processes, downloading and uploading mp3 files. Thus, anyone may potentially contribute an agent (or more) to the society, but before it joins the society it is registered at the central server.

We argue that semi-open societies only slightly limits the openness compared to completely open societies, but have a much larger potential for providing stability and trustfulness. For instance, it is possible to monitor which agents are currently in the society. This also makes the boundary of the society explicit.

4 Semi-closed Agent Societies

In what we will refer to as semi-closed societies, external agents are not allowed to enter. However, they have the possibility to initiate a new agent in the society which will act on the behalf of the external agent. This is done by contacting a kind of institution representing the society and ask for the creation of an agent (of one of a number of predefined types). The institution then creates an agent of this type, which enters the society with the goal of achieving the goals defined by the external agent. (See Figure 2.)

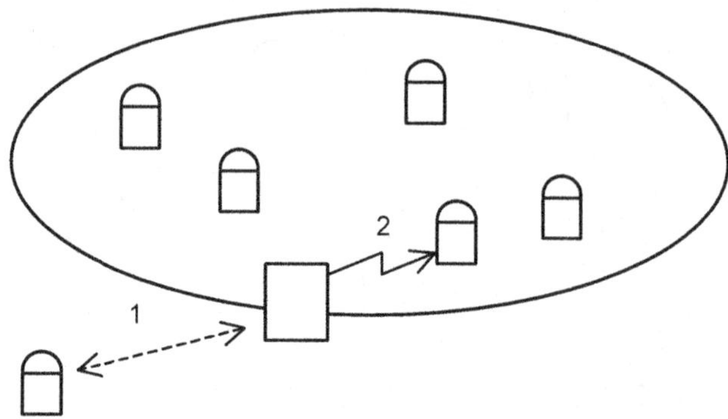

Fig. 2: An Agent Initiating a Representative in a Semi-closed Society.

All agents are run on the same (set of) servers. Typically, the agents are implemented and the servers are managed by a third party, i.e., the owner of the society. As the possible behaviors of all agents are known, it is easier to control the activities in the society.

The following example of a semi-closed society is based on the activity of searching for and booking "last-minute" holiday travel tickets. The price of this type of tickets may change on a daily basis and are determined by current supply and demand. For instance, if the whether is nice at the place from where the travel departures, prices on travel tickets to Mediterranean beach resorts typically drop as the departure date approaches. Today, customers find and buy tickets manually by browsing newspaper ads and WWW-pages, as well as phoning and visiting local travel agents. Most people regard this as a time-consuming and boring activity and would probably be happy to let software agents do the job.

The society is based on an existing prototype implementation.[1] Customers specify their preferences (departure date, destination, max. price etc.) through either a WAP or WWW interface. An agent is then initiated on a service portal (at a remote computer) with the goal of finding a ticket satisfying the customer preferences. It continually searches a number of databases until it reaches its goal (or is terminated by the customer). When a ticket is found, the agent either books it directly, or sends an SMS to the customer asking for confirmation. To book the ticket, the customer agent contacts an agent representing the travel company (the info in the database contains the address to the travel agent). When the customer agent receives a confirmation, it immediately sends an SMS message to the customer about this and then terminates.

---

[1] The system was developed during the first five months of 2000 by a group of 14 students doing their final exam project for their B.Sc. theses at Blekinge Institute of Technology, under the supervision of the author. It was a joint project with HP, who donated servers, workstations, and E-speak licences, and Aspiro, a Swedish company in the area of Internet-based services, who contributed knowledge about development of Internet-based services and reuse components. E-speak is a platform for service management developed by HP and was used for agent administration and communication.

The description above corresponds to the current version of the system. However, it is easy to imagine possible future extensions, e.g.:

- letting the customer agent and the travel company agent negotiate about the price,
- auctioning off the remaining tickets when there is only a specified number of hours or days left until departure,
- letting the travel company agents autonomously decide the ticket price based on available information about the current situation (e.g., supply, demand, and number of days before departure).

Semi-closed societies provide almost the same degree of openness as semi-open societies but are less flexible. On the other hand, they have a larger potential for implementing stability and trustfulness. An interesting property of the semi-closed societies is that they seem to indicate the limit of how open a society could be where the owner of the society may still control the overall architecture of the society. To have control over the architecture is a prerequisite for applying many of the ideas on how to achieve multi-agent coordination (cf. Lesser [7]). Moreover, this type of society poses interesting questions regarding ownership: Who is actually responsible for the actions of the agents?

5 Conclusions

We have described four different categories of artificial societies that balance the trade-off between openness, flexibility, stability, and trustfulness differently. Based on the analysis of completely open and completely closed societies, which revealed that open societies support openness and flexibility but not stability and trustfulness and that the opposite is true for closed societies, we suggested two other categories, namely semi-open and semi-closed societies. We argue that these types of society let us have the best from both worlds. Whereas semi-open societies are more flexible than semi-closed societies, they have lower potential to achieve stability and trustfulness. A summary comparison between the different types of societies (ordered in degree of openness) is provided in Table 1.

The balancing of trade-off provided by the semi-open and semi-closed societies is necessary for implementing systems that can be characterized as information ecosystems. In this type of systems there is a strong need for mechanisms for “enforcing” ethical behavior between agents in order to provide trustful systems to end-users. In completely open societies such mechanisms probably need to be very complex (if they exist at all), which means that the potential for achieving trustful systems is very low. In completely closed systems, on the other hand, the potential for achieving trustfulness is great, but the price you have to pay, by making it impossible for new agents/owners to enter the society, is too big in applications where openness is desired.

Table 1. A Comparison between the Different Types of Artificial Societies.

	Fixed	*Closed*	*Semi-closed*	*Semi-open*	*Open*	*Anarchic*
Agents	Fixed at design time	Known at design time	Known at run time	Known at run time	Cannot be known	Cannot be known
Constraints	Fixed	Fixed	Fixed	Fixed	Not fixed	Not fixed
Communication language	Fixed	Fixed	Fixed	Fixed	Usually fixed	Not fixed
Roles	Fixed	Fixed	Fixed	Usually fixed	Usually fixed	Not fixed
State	Can be monitored	Can be monitored	Can be monitored	Cannot be monitored	Cannot be monitored	Cannot be monitored
Agent owners	Fixed at design time	Fixed at design time	Can be known	Can be known	Cannot be known	Cannot be known
Society owner	yes	yes	yes	yes	no	no

5.1 Future Work

In this work we focussed on four properties of artificial societies that we believed were the most important. However, there are a number other properties that may be relevant in some domains. Examples of such properties are: *fairness*, i.e., the degree to which the society members are treated equally; *performance*, i.e., how efficient is the society (measured in e.g., time or number of messages needed to perform a certain task), and *"controllability"*, i.e., the support for the owner of the society to control the activities in the society, e.g., by punishment. Future work will take also take these and other relevant aspects into account.

An important part of the future work is the definition of appropriate metrics for quantifying the properties of artificial societies and developing methods for measuring them. Also, methods for determining how well different types of artificial societies balance the trade-off between these aspects need to be developed. Such methods could be based on simulation experiments and/or theoretical analyses.

Finally, both theoretical and practical aspects of institutions need further investigation. The theoretical aspects include questions regarding what functionality actually is possible to implement and what is not. The practical work includes the development of mechanisms for: accepting new agents to a society, monitoring the behavior of individual agents, and "enforcing" ethical behavior between agents in order to provide trustful systems to end-users.

Acknowledgements

The author acknowledges the valuable contribution from the colleagues in the ALFEBIITE project (IST-1999-29003) and the members of the Societies of Computation research group at Blekinge Institute of Technology.

References

1. A. Artikis and J. Pitt. A Formal Model of Open Agent Societies, In Proceedings of the *Fifth International Conference on Autonomous Agents*, 2001.
2. K.M. Carley and L. Gasser. Computational Organization Theory. In G. Weiss (editor), *Multiagent Systems*, MIT Press, 1999.
3. P. Davidsson. Emergent Societies of Information Agents, In *Cooperative Information Agents IV*, Springer Verlag, LNCS 1860, 2000.
4. J. Ferber and O. Gutknecht. A Meta-model for the Analysis and Design of Organizations in Multi-Agent Systems. In Proceedings of the *Third International Conference on Multi-Agent Systems*, IEEE Computer Society, 1998.
5. R. Gustavsson and M. Fredriksson. Coordination and Control of Computational Ecosystems: A Vision of the Future. In: A. Omicini, M. Klusch, F. Zambonelli, and R. Tolksdorf, editors, *Coordination of Internet Agents: Models, Technologies, and Applications.* Springer Verlag, 2001.
6. S.J. Johansson and J. Kummeneje. A Preference-Driven Approach to Agent Systems. To appear in the *Proceedings of the Second International Conference on Intelligent Agent Technologies*, 2001.
7. V.R. Lesser. Reflections on the Nature of Multi-Agent Coordination and Its Implications for an Agent Architecture. *Autonomous Agents and Multi-Agent Systems*, Vol. 1: 89-111, Kluwer, 1998.
8. A. Oram (editor)*Peer-to-Peer: Harnessing the Power of Disruptive Technologies* , O'Reilly, 2001.
9. M.P. Singh, A.S. Rao, and M.P. Georgeff. Formal Models in DAI, In G. Weiss (editor), *Multiagent Systems*, MIT Press, 1999.
10. F. Zambonelli, N.R. Jennings, and M. Wooldridge. Organizational Abstractions for the Analysis and Design of Multi-Agent Systems. In P. Ciancarini and M. Wooldridge, editors, *Agent-Oriented Software Engineering*. Springer Verlag, LNCS 1957, 2001.

A Methodological Perspective on Engineering of Agent Societies

Martin Fredriksson and Rune Gustavsson

Department of Software Engineering and Computer Science
Blekinge Institute of Technology
S-372 25 Ronneby, Sweden
{martin.fredriksson,rune.gustavsson}@ipd.bth.se

Abstract. We propose a new methodological approach for engineering of agent societies. This is needed due to the emergence of the Embedded Internet. We argue that such communication platforms call for a methodology that focuses on the concept of open computational systems, grounded in general system theory, and natural systems from an engineering perspective. In doing so, it stands clear that forthcoming research in this problem domain initially have to focus on cognitive primitives, rather than domain specific interaction protocols, in construction of agent societies.

1 Introduction

Communication and information technology can be considered to be at the forefront of technological advancement with high societal impact. Evidence of this has manifested itself in the successful acceptance of stationary communication platforms, such as the Internet, and virtually everywhere in our societal infrastructures and social activities. In addition to this, the development and deployment of mobile communication platforms has also gained momentum. As a result from combining these communication platforms we get a merged communication network, commonly denoted as the *Embedded Internet* [4]. The main features and related issues of the Embedded Internet have previously been identified by Tennenhouse in terms of *proactive computing*, where he describes the network, growing with a factor of eight billion new interactive computational nodes per year, to be in serious need for a unified approach to construction and observation of the network's behavior:

> *"One can imagine the underpinning of computer science taking a leap from deterministic to probabilistic models in much the way that physics moved from classical to quantum mechanics. ... Very little has been done by way of adopting lessons learned in the mainstream computer science systems and theory communities or in rethinking the fundamentals of the various engineering disciplines in light of cheap, plentiful, and networked computation."* – Tennenhouse, D. [15].

From a qualitative perspective, the emergence of communication platforms such as the Embedded Internet obviously introduces us to many new, interesting and unique opportunities. However, along with the opportunities comes an equal amount of problematic

A. Omicini, P. Petta, and R. Tolksdorf (Eds.): ESAW 2001, LNAI 2203, pp. 10–24, 2001.

issues that have to be dealt with up front. Tennenhouse identifies three primary areas of investigation concerning the behavior of proactive systems (e.g. the Embedded Internet): (i) pervasive coupling of networked systems to their environments, (ii) bridging the gap between control theory and computer science, and (iii) proactive modes of operation in which humans are *above the loop*. Currently, much of the research addressing these issues does so in terms of distributed artificial intelligence (multiagent systems) and software engineering. In addressing the emergent issues as described by Tennenhouse, a common approach is to assimilate conceptions and metaphors related to the notion of system organization and control, mainly from the area of sociology. However, as argued by Malsch in *Naming the unnameable* [11], there is a danger in such an approach since it more often than not focus on construction of single agents, modeled in terms of sociological metaphors. It is important to understand that the sociological answers we seek, e.g. organization and control, will not be found as long as the idea of *single agents* with predetermined motives, interests, and intentions prevails. Previously, this argument was advocated by Gasser in a notable passus of his:

> *"... DAI systems, as they involve multiple agents, are social in character; there are properties of DAI systems that will not be derivable or representable solely on the basis of properties of their component agents. We need to begin to think through and articulate the bases of knowledge and action for DAI in the light of their social character."* – Gasser, L. [5].

The general message brought to light by such statements can be considered as twofold. First, we must necessarily avoid making use of metaphors in construction of complex artificial systems, if the constructed behavior of system components does not conform to the behavioral principles, or *laws*, of the theoretical understanding and foundations of the metaphors. Secondly, when we consider open systems, artificial or not, these must necessarily be grounded in the notion of behavior as a result from interaction between multiple agents, as opposed to internal behavior of a single agent. Consequently, the material presented in this paper is focused on three key statements of ours. We claim that we have to develop new engineering methodologies that, from a general perspective on open and proactive systems, focus on building, deploying and maintaining new services of the Embedded Internet. We also claim that multiagent systems will be one of the enabling technologies utilized by those new methodologies. Finally, we claim that identification of new appropriate methodologies have to begin with assessments of underlying (hidden) assumptions of current software engineering paradigms as well as assessments of principles behind other (civil) engineering paradigms and models of complex systems. The paper is organized as follows. In the next section we set the stage by making a comparison of different engineering principles. We propose new engineering principles of software that are in accordance with other engineering principles with one important exception. We have to engineer the *nature and societal fabric* of computational systems as well. Following that line of reasoning we introduce basic *World Views of Multiagent Systems* in Section 3 and *The World View of Open Systems* in Section 4. The synthesis of our work is summarized in Section 6 on *Engineering of Agent Societies*. We conclude the paper, in Section 7, with a summary and pointers to suggestions for future investigations and research.

2 Comparison of Different Engineering Principles

A major part of the currently deployed computational systems are based on an existing foundation of embedded and communicating systems that are inherently distributed with respect to location as well as control and access. As described by Gustavsson et al., yet another aspect of these systems is their focus on providing services of a qualitative nature to various users in a societal setting [8]. Therefore, in many systems of this nature, properties such as added customer value and trust are of key importance in order for system acceptance to manifest itself. The challenge confronting future system developers can be formulated as follows. There will be no such thing as system development in the traditional meaning. That is, few systems (applications) will be developed from scratch, we will rather focus on enhancements of existing software by including new functions and services based on assessments of existing system capabilities and constraints.

2.1 Models of Society and Nature

In this respect software engineering will be more in line with traditional engineering. We are, for instance, building new communication vehicles respecting societal laws (safety, pollution and so on) as well as natural laws (laws of friction, physical and chemical laws of combustion engines and so on). Automobile engineering is about building artefacts that can be deployed or retracted from the *existing* open society of surface-based communication. The automobile engineering can thus focus on added value for customers or benefits for the society, such as safety and less pollution, *as long as* the society and natural laws are respected. The last sentence is a key statement. It presupposes that those laws and norms, of the (conditional) open society of surface-based communication, are *known* to the engineering process. Hence, the main difference between scientific research and engineering in the natural sciences is that scientific research focuses on understanding aspects of our natural environment by finding models and laws that capture relevant aspects. In engineering, those scientific models are tailored into well-established engineering principles, allowing us to build artefacts with intended functionality and behavior. In traditional engineering, nature has the final saying. It is impossible to build an artefact whose functionality or behavior requires that some natural law is violated. If we related this line of reasoning to observation and construction of open computational systems, it should be noted that we have come to the point where we actually have to come to terms with the foundations of behavior in open systems. Either we continue to create isolated artifacts, that by appearance remind us of their physical equivalence, e.g. humans versus intelligent agents, or we focus on artifact interactions that are grounded in principles of natural systems. Following this chain of reasoning we firstly state that the Embedded Internet is, or will soon be, of such complexity that we can and must regard it as a physically grounded environment. Secondly we have to recast engineering of software accordingly, that is; software engineering has to focus on development of computational entities that are respecting societal and natural *laws of interaction.*

2.2 Models of Computation

In light of the previous line of reasoning, engineering of a sustainable Embedded Internet is much more complex than traditional engineering of physical artefacts. That is, we have to develop *both* nature *and* corresponding laws in parallel and in accordance with the existing infrastructure. To do so, we have to begin with an assessment of some basic underpinnings of current paradigms of scientific research in computing and in software engineering. Scientific research in computing is focusing on investigations of models of computing. Examples include programming of abstract machines, leading to theories of algorithms, protocols, automata and formal languages, complexity theory, verification of correctness of programs, algorithms, protocols, etc. In short, scientific research in computing has been focusing on setting up models of computations and verifying properties of such models. This line of research is in fact a field of applied mathematics. Such a methodological approach to scientific research in multiagent systems can, for instance, be found in research on agent communication languages, as described by Wooldridge [17]. The theoretical findings have then led to better and more efficient engineering principles of software, e.g. parsers, compilers, and debugging tools.

2.3 Coupling of Methodological Approaches

We can summarize scientific research and engineering paradigms in *natural systems* as a bottom up phase, e.g. modeling of relevant aspects of the existing reality, followed by a top down phase where scientific findings of the models are transformed into well-established principles of engineering. We argue that the current coupling between scientific research and engineering in computing only takes into account the top-down approach. Up to this point, such a methodological approach has been sufficient due to the focus on standalone application development, i.e. engineering of closed computational systems. We argue that in order to meet the challenges of application development in the future Embedded Internet, we also have to include a bottom-up phase in our methodology. In many cases an appropriate methodological approach to systems engineering is indeed to primarily consider the physical implementations of some model rather than the abstract computational model. Examples of such qualitative approaches to engineering of multiagent systems and agent societies have previously been advocated by Gustavsson et al. [7] and Castelfranchi [2]. In order to develop tools for a bottom-up modeling approach of the existing Embedded Internet, we suggest an assessment of methodologies developed in system theories of natural and artificial systems. The purpose of such an approach is to observe and construct (discover, design, and implement) an appropriate set of *societal and natural laws* of our system, e.g. the Embedded Internet. We return to these topics in Section 4 on *The World View of Open Systems* and in Section 6 on *Engineering of Agent Societies*. However, as a preparation, we start with a proposal for world views of multiagent systems in the next section.

3 World Views of Multiagent Systems

It is often the case that we make a distinction between two unique tools used in research activities: the scientific method as such and the methodological approaches put to use

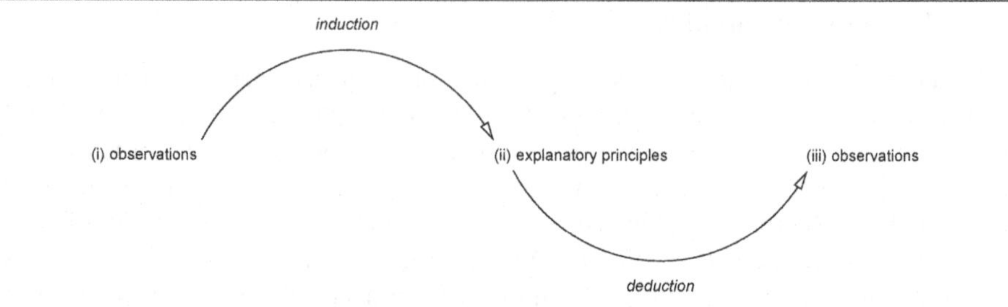

Fig. 1. The inductive–deductive procedure of the scientific method. An explanatory inquiry begins with knowledge that certain events occur. Explanation is achieved only when statements about these events are deduced from (ii) explanatory principles. Hence, explanation is an inductive–deductive transition from (i) knowledge of an event to (iii) knowledge of the cause for the event.

in applying the scientific method. We consider the scientific method (Fig. 1) to be the full spectrum of activities ranging over: (i) observation of phenomena, (ii) formulation of hypotheses, (iii) prediction of new observations, and (iv) performance and analysis of experimental tests. As a complement to the scientific method and its involved activities we consider different methodological approaches. The intent of applying these approaches is obviously manyfold, but one aspect is that they should help us in describing phenomena observations and corresponding behavioral predictions. Consequently, a very important component of any methodological approach is something often denoted as the world view. However, before delving into the currently applied methodological approaches in engineering of agent societies, let us try to clarify the general concept of a methodological approach. Note that the particular approach described in this paper addresses man made systems, i.e. systems that (at first) are not *a priori* part of nature, but rather introduced by means of human intervention. Still, as a result from the systems being of an open nature, i.e. subjects to external stimuli, their behavior are to a great extent under the influence of their respective environments.

3.1 A Methodological Perspective

An example of a methodological approach towards engineering of (knowledge-based) information systems was introduced by Schreiber et al.[14] in knowledge engineering, involving the following components: world view, theory, methods, tools, and usage.

- *World View*. The world view component explicitly defines the boundaries of a certain problem domain, in terms of structural models and their constituents. In principle, these models define what entities supposedly exist in a system and in what way they relate to each other. Hence, an example of a world view is consequently what we often refer to as an ontology.
- *Theory*. Based on the world view component, a hypothesis can be constructed. The main intent of such a hypothesis is to assert or refute certain theories concerning

explanatory principles of the system considered in the world view. The theory component of a methodological approach therefore involves the construction of theories concerning system behavior.

- *Methods and Tools.* Grounded in the world view and its corresponding theory component, a number of methods and tools can be identified that seek to aid the activities of observing a system. Consequently, the methods and tools derived from the world view component and the theory component seek to aid us in deducing new observations from a prior set of observations in conjunction with certain explanatory principles.
- *Usage.* As a result from the previous components of the methodological approach, we have a set of models at our hands, describing certain features and aspects of a given system, that will be put to some specific use. Hence, it should be pointed out that a model can be used as input to two completely different activities: verification and validation.

Obviously, it is important to emphasize the strong connection between world view (observation) and the problem domain of model usage (observation and construction). From a pragmatic perspective, it can be said that no matter what type of system or specific phenomenon we are currently dealing with, it must primarily be possible to both observe and construct such a system using the same set of models. If this is not the case, there must necessarily be something wrong with our fundamental world view, being the basis for all model creation. Consequently, it should be pointed out that this methodological perspective is based on the assumption that in principle there is no difference between the behavior of physical and natural systems. We consider the behavior of natural systems to be spontaneous, or emergent, and we consider the behavior of physical systems to be constructed, or designed, from well known principles. Both types of systems manifest themselves in terms of processes and structural evolvement, but the former is not primarily considered to be man made. Examples of such system classes are, respectively, chemical compounds (natural system) and multiagent systems (physical system).

3.2 A Multiagent Systems Perspective

Distributed computing, as manifested in increasingly sophisticated net-based applications, is an area of major importance in computer science and software engineering, but also to our society at large. With the advent of the Embedded Internet, in combination with previous research on artificial intelligence, organization, and control, we are currently addressing advanced distributed applications in terms of multiagent systems construction and control. The main reason for this is probably that such methodological approaches explicitly address complexity issues in computational systems that necessarily cannot be considered to function in isolation, but rather under the influence of stimuli from the surrounding environment:

> *"... intelligent systems do not function in isolation – they are at the very least a part of the environment in which they operate, and the environment typically contains other such intelligent systems."* – Huhns, M. N. [9].

As a result from the fact that the environment under study is situated in a societal context, identification of different perspectives on the nature of these societies of agents has started to emerge. From one point of view, the construction of agent societies is of major concern, and a focus on technical aspects of system construction is introduced. An example of this (physical systems) perspective is described by Jennings [10], in that issues of system complexity is addressed in terms of decomposition, abstraction, and organization of multiagent systems. Yet another point of view is that of control and organization and, since the system under study is situated in a societal context, an explicit focus on qualitative aspects of system organization is introduced. A notable example of this perspective has previously been introduced by Castelfranchi [2], in that issues of system complexity is addressed in terms of emergent behavior as a result from local design of interacting agents:

> *"... obtaining from local design and programming, and from local actions, interests, and views, some desirable and relatively predictable / stable emergent results."* – Castelfranchi, C. [2].

If we consider the existing Embedded Internet as a basic starting point for our investigation of multiagent systems in societal settings, we necessarily have to come up with a common conceptual framework that deals with both observation of emergent system behavior as well as engineering of the involved system structures. In pursuit of such a common conceptual framework, we believe that general system theory and the concept of *open computational systems* is an appropriate starting point. The concept of open systems as such has taken on a prominent role in research disciplines dealing with computational complexity, e.g. multiagent systems and software engineering, and consequently, issues related to system organization and construction are addressed up front in these areas of research. Unfortunately, there is little consensus concerning a common understanding of what an open system is. Therefore, in the next section of this paper we will try to clarify the concept of open systems as it is currently applied in a range of scientific disciplines. These perspectives, or rather their differences, are then summarized in order to present an initial understanding of open systems in our context.

4 The World View of Open Systems

From Popper's point of view, open systems are considered as systems in a state, far from equilibrium, that shows no tendency towards an increase in disorder [13]. Such structural disorder in a system is described in terms of entropy; corresponding to a measurement of some subject's structural organization. Initially, the notion of entropy stems from the second law of thermodynamics; stating that the molecular disorder of a closed system can only increase until it reaches its maximum, i.e. total disorder. In the more general case of an open system, entropy is considered as possible to export into the surrounding environment and, hence, the system's structural order can increase rather than decrease. Consequently, from the perspective of organization, an open system can develop structural properties that are the very opposite of turning into an equilibrium state. Nicolis and Prigogine describes the development of such far from equilibrium properties as attributed to a flow of mass or energy in the system, to or from a system's surroundings [12]. In effect, it is primarily the origin and destination of this flow that classifies a system as open or closed (external versus internal). Another perspective of open systems is that which is found in various research disciplines related to computer science and computational complexity, e.g. multiagent systems and software engineering. The concept of an open system is then primarily considered in terms of a dynamic set of interacting entities, where no *single* individual or organization is in control of the construction or, consequently, behavior of the set as a whole. One example of such a (societal) perspective on open multiagent systems has been advocated by Davidsson [3]. However, the interactions that take place between the entities in the set are not of an open nature, in the sense that each interaction necessarily must be constructed according to some syntactical structures, i.e. an interaction protocol, agreed upon prior to the construction of the involved entities. Consequently, an open system from the latter perspective could be defined as a dynamic set of interacting entities, restricted by the precedence of interaction protocols over entity construction. Cf. our previous discussion in Section 2 on restrictions imposed on a system by natural and societal laws that exist *a priori*. Obviously, the difference between definitions on the concept of open systems lies in the very nature of the systems under study. The common denominator of such systems is their inherent complexity. However, the systems can be said to differ in respect of their origin. The first perspective deals with system complexes that primarily are *a priori* a part of nature. The second perspective, on the other hand, deals with systems of an abstract nature, i.e. the involved complexes are not *a priori* a part of nature. In effect, if we are interested in broadening our understanding of system construction and engineering as well as observation of emergent system behavior, the common conceptual framework of open computational systems must necessarily address system properties from a more general systems perspective.

4.1 A General Systems Perspective

As previously described, one of the main intents of this paper is to identify a proper starting point regarding the definition of a conceptual framework for open computational systems. A basic condition for the existence of such systems is considered to be that of the Embedded Internet. In an open computational system there is no single individual or organization that is responsible for the construction and consequently the organization of the system as a whole. Therefore, if we are interested in systemic properties of the system as a whole, we have to make use of a conceptual framework that enables us to generalize systemic properties of an open system's parts into an understanding of systemic properties of the system as a whole. Such a theoretical framework has previously been advocated in the form of general system theory:

> *"Thus, there exist models, principles, and laws that apply to generalized systems or their subclasses, irrespective of their particular kind, the nature of their component elements, and the relations or forces between them. It seems legitimate to ask for a theory, not of systems of a more or less special kind, but of universal principles applying to systems in general."* – von Bertalanffy, L. [16].

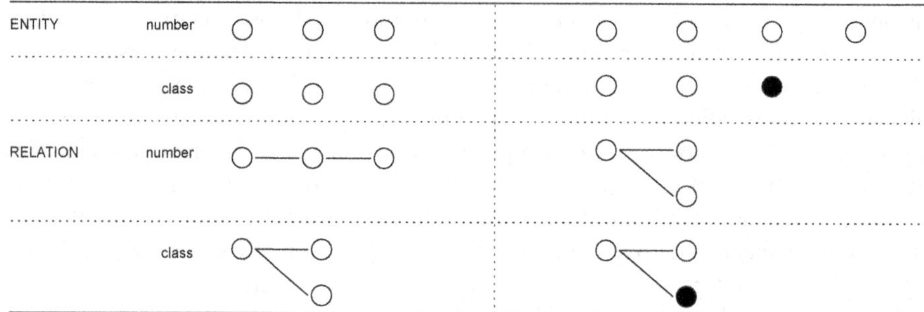

Fig. 2. An illustration of system primitives of an open computational system. Both entities and relations can differ in number and/or class.

General system theory is primarily grounded in the notion that the whole is more than the sum of its parts. The meaning of this notion is simply that the constitutive characteristics of a given system, or complex, are not solely explainable from the characteristics of isolated parts. In effect, a system's constitutive characteristics therefore appear as new or emergent. However, if we know the total of parts and the relations between them in a system, the global behavior of the system can be derived from the behavior of the individual parts. From a holistic perspective this means that if the interactions between a system's parts cannot be considered as non-existent (e.g. weak enough to be neglected) or that the relations describing the behavior of parts are not linear, we necessarily must take on a constitutive, rather than summative, approach in describing the behavior of the system. In general system theory, a system is described in terms of complexes of interacting entities. The entities as such can be considered in isolation from three different perspectives: number, species, and relations. These different perspectives of the complexes constituting a system are the basis for two classes of characteristics: summative and constitutive. Summative system characteristics are those that may be obtained by means of summation, within or outside the complex, e.g. number of entities. Constitutive system characteristics on the other hand are characteristics that depend on specific relations within the complex, i.e. we must know not only the parts, but also their relations. In light of the previous discussion on similarities and differences between the scientific disciplines and our understanding of general system theory, let us define an open computational system more precisely.

4.2 An Open Computational Systems Perspective

In summary, we have understood that a system is considered to be a dynamic set of entities that interact. Furthermore, a system can only be characterized as open in relation to certain system properties, e.g. natural and societal laws as described in Section 2. According to general system theory, this leaves us with three primitive properties of any given system that are affected by such laws: number of entities, entity species, and entity relations. However, we argue that it is not only the entities that belong to some species, or types, but also the relations (Fig. 2). In order to deal with this issue, we characterize the primitive properties of an open computational system as *entities* and *relations*. This introduces us to the system property that renders a system as open, namely its physical structure. In principle, this means that we no longer consider an open computational system to primarily be the manifestation of an abstract model, but rather that it is grounded in the physical world and that we seek a conceptual model that helps us to understand the system's emergent properties.

– *Entities*. As with all systems, a computational system can be modeled and analyzed at different levels, or layers, of abstraction. These different layers all share a common property, namely that they can be described as constituted by a set of entities that in some way are related to each other. Hence, we consider any system to be a population of an unknown number of physical entities. We denote such physical entities as *primitives*. From a cognitive perspective, an entity can also be a combination of two or more primitive entities in relation to each other. We denote such entities as *composites*. As a result from this perspective of abstraction, each composite can be described as adhering to a certain class, primarily characterized by its structure.
– *Relations*. Given that we are studying systems of a physically open nature, all entities can interact with each other by means of a number of relations. In line with the previous discussion on cognitive properties of entity composition, the relations of an entity can also be considered in terms of primitives and composites. In other words, if there exists a relation between two primitive entities, we denote such a relation as *primitive*. Consequently, if there exists a relation between two composite entities, we denote such a relation as *composite*. From a cognitive perspective, a composite of relations can be described as adhering to a certain class, primarily characterized by its structure.

In light of the previously described world view of open computational systems, any entity can be considered as a distinct and separate existence that from a cognitive perspective is perceivable by some external party. Along with this description comes the fact that all entities can have either direct or indirect relations to each other. This focus on primitive and composite structural properties of a natural system and its constituents introduces us to a quite important issue:

> *"Because the entity exists independent of all other things, its behavior is known only to itself. Unless the entity shares its behavior with a different entity, no one has knowledge of its unique behavior."* – Ghosh, S. [6].

5 The Theory of Open Systems

The previously described world view of open computational systems identifies an important difference in our understanding of construction versus observation of system behavior. If a system can be characterized as open, this must necessarily mean that no single individual or organization is solely responsible for its construction, and consequently, that no single individual or organization can be fully aware of its true behavior either. Such a property of a system is what we would refer to as emergent behavior.

Hence, an important question we should ask ourselves is therefore what systemic properties we actually can observe if we are dealing with open computational systems. In search for an answer to this question, we argue that our understanding of construction and observation of open computational systems could benefit from a dynamical and complex systems perspective.

5.1 A Dynamical Systems Perspective

From a pragmatic perspective, an entity does not necessarily have an explicit relation to all other entities in its surrounding environment, but from a systemic perspective all entities actually do have some form of direct or indirect relation with all entities in the surrounding environment. These entities and relations form the structure of a system. In software engineering this structure is considered to be of a closed nature and hence local coherence of the system is in focus. That is, the focus is on a specific application where the physical components and their interaction in principle is a stable result from the design-phase of the application development process. However, with open computational systems, it is global coherence that is of the essence. The notion of global coherence focuses on the idea that a system's structure evolves over time as a consequence of communication between the involved entities. The description of such structural evolvement is called the behavior of a system, and in computer science these issues are addressed in research and development on coordination, control, and organization. This dynamical systems perspective is addressed in terms of structures, patterns, and processes.

- *Structures.* The basic property of any given system is its structure, as characterized by an unknown number of physical entities and their relations. However, it is not structures that are subject to principle investigation in dynamical and complex systems but rather patterns and the evolvement of these patterns through certain processes.
- *Patterns.* A pattern in dynamical systems is considered to be the abstract manifestation of the system's physical structure, or configuration of relationships [1], and it is such patterns that are considered to form the basis for emergent behavior of some system. Hence, if we are interested in system dynamics, we are primarily interested in studying patterns and their evolvement over time, presupposing that they are abstract derivations from some natural or physical system.
- *Processes.* When it comes to system dynamics, a process can be considered as the continuous manifestation of some system's pattern of organization. In principle, this means that processes are, in the same way as patterns, of an abstract nature, derived from an underlying natural or physical structure.

An example of system dynamics in terms of the previously described concepts would be that of a feedback loop in a societal setting. Generally speaking, a feedback loop can be defined as a cyclic path in a network pattern, where a signal travels along some path in the underlying physical structure and visits all nodes in such a way that after the signal has visited the last node, the signal returns to the first node. This process continues until the pattern of organization is changed in some way. The derivation of such a process, e.g. the invisible hand in social sciences [2], can be said to correspond to the self-organization of some system's physical structure. For example, a negative rumor, concerning the performance of some entity, is introduced into a cyclic path of communication between a set of entities involved in the trade of some commodity. As a result of this rumor, the reputation of the implied entity is spread throughout the network in the form of a negative reputation, resulting in the exclusion of the entity. In effect, the structure, as well as the corresponding pattern, is reduced by one node and the system's organization has changed.

6 Engineering of Agent Societies

When it comes to engineering of agent societies, i.e. the Embedded Internet, we state that there are currently two distinct approaches applied in this problem domain: construction of system structures and observation of system behavior. Obviously, there is quite a profound difference in focus between the approaches, but we believe that they share a common (implicit) denominator; an open computational systems perspective on the environment under study. Previously in this paper, we have therefore presented an initial understanding of the general assumptions and implications of the two approaches (see Section 2 *Comparison of Different Engineering Principles* and Section 3 *World Views of Multiagent Systems*). In summary, we argue that, with the advent of the Embedded Internet, construction and observation of agent societies should no longer be considered as separate disciplines, nor should the subject of investigation be considered in terms of metaphors, but rather as grounded in principles of interaction laws in natural systems. A methodological approach to engineering of agent societies must necessarily comprise principles of *both* observation *and* construction that applies to the whole system. The reason for this is primarily that *all* constituents of the system under study are able to communicate with each other. Consequently, in this section we outline the basic structure of such a methodological approach, that explicitly focuses on an open computational perspective on engineering of agent societies.

6.1 An Open Computational Systems Approach

The foremost difference between construction and observation of agent societies lies in the relation between system and model, as described in Section 2. If we explicitly focus on construction of a system, it is implicitly stated that the models have precedence over the system, in that they are used as templates for system behavior. Such an approach also implies that models can be nothing but subject to verification, since the process of validation necessarily assumes that there exists some corresponding real system *a priori*. This does not necessarily pose a problem, if we would be dealing with systems of a closed nature. However, if we acknowledge the fact that new species of agents possibly could enter a system at any point in time, rendering it as open, we can no longer be certain of a system's behavior at design time. If not always the case, this situation is very common in the physical sciences; where the system under study is considered to have precedence over the corresponding models. Alas, prior to construction and deployment of an agent into some system we must first observe and establish general principles of

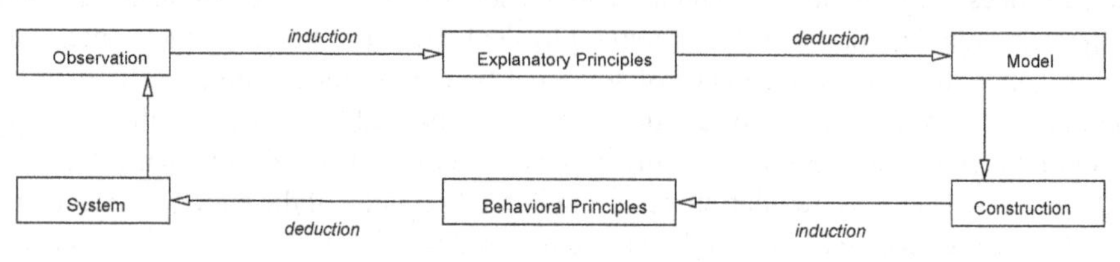

Fig. 3. The general structure of a methodology for engineering of agent societies. Observation of a system, existing *a priori* in nature, results in a model from applying a set of explanatory principles to observed system properties. Construction and deployment of agents is then performed by applying certain behavioral principles to a given model.

system behavior that might affect the agent's behavior. This activity of observing the general system properties of an open computational system could be, as argued in Section 2, described as the identification of natural and societal laws of agent societies. In summary, we state that if *all* entities in a system at any point in time are able to communicate with each other, as is the case with the Embedded Internet, we must necessarily consider a methodological approach that comprises both principles of system observation as well as those of system construction (Fig. 3). From a general perspective, the methodology we propose is constituted by the traditional approach of the inductive–deductive procedure, coupled with the principles of any engineering procedure. Furthermore, it should be noted that we assume that the term *system* in this context refers to a universal and open system and not some closed subsystem in isolation from its environment. Given this general methodological perspective, it stands clear that we seek an approach to engineering of agent societies that offers us the possibility to deduce and induce systemic properties of a general nature. Hence, the foremost requirement imposed on the methodological approach that we propose is that any part(s) of the open computational system must be possible to model using the same general systemic primitives. We deal with this requirement in terms of *entity complexes* and corresponding *relation complexes*. Furthermore, the structures formed by these complexes are considered as the fundamental basis in modeling of system dynamics, in terms of *patterns* and *processes*. Consequently, the methodological approach we propose calls for methods and tools in engineering of agent societies that reflects the previously described modeling primitives and principles of behavior in natural systems. Not until then will we be able to observe or construct the behavior of the Embedded Internet as a whole system. The foremost utility of our proposed methodological approach to engineering of agent societies is that, by focusing on general system primitives and principles, we can directly apply the models resulting from explanatory principles in observation to the behavioral principles in construction.

7 Discussion and Concluding Remarks

We have outlined a methodological approach in terms of a corresponding world view and an initial understanding of its theoretical foundations. In summary, our methodological approach identifies an open computational systems perspective to be the basis for a common world view. This world view deals with structural properties of a system in terms of entities and relations from a natural systems perspective. However, since the world view of any methodological approach typically only addresses cognitive properties of the system under study, we have also proposed an initial theory that focuses on system dynamics and behavior. This theory is based upon the dynamical systems perspective on behavior of open systems, also grounded in a natural systems perspective on engineering. Following this line of reasoning, it stands clear that forthcoming research in this problem domain initially has to focus on methods and tools in construction that reflects the requirements and needs of observation and explanation of natural system properties. Not until then can we understand in what way the deployment of agents into our societal fabric is affected by the *a priori* existing system, or in what way the system itself is affected by the introduction of new and unknown agents. In an attempt to achieve this, we are currently conducting a project at the Societies of Computation Laboratory called SOLACE (Service-Oriented Layered Architecture for Communicating Entities). The main objective of the project is to study mechanisms that are involved in the continuous manifestation of systemic patterns in open computational environments, grounded in principles of behavior in natural systems.

Acknowledgments

The authors wish to thank all members of the Societies of Computation research group at Blekinge Institute of Technology, for stimulating discussions and useful comments on earlier drafts of the paper and related material. The authors also acknowledge the valuable contribution from colleagues in the ALFEBIITE and SOLACE projects.

References

1. F. Capra. A new synthesis. In *The web of life: A new scientific understanding of living systems*, pages 157–176. Anchor Books, 1996.
2. C. Castelfranchi. Engineering social order. In A. Omicini, R. Tolksdorf, and F. Zambonelli, editors, *Engineering societies in the agents world*, volume 1972 of *Lecture notes in computer science*, pages 1–18. Springer, 2000.
3. P. Davidsson. Emergent societies of information agents. In M. Klusch and L. Kerschberg, editors, *Cooperative information agents IV: The future of information agents*, pages 143–153. Kluwer Academic Publishers, 2000.
4. D. Estrin, R. Govindan, and J. S. Heidemann. Embedding the internet: Introduction. *Communications of the ACM*, 43(5):38–41, 2000.
5. L. Gasser. Social conceptions of knowledge and action: Dai foundations and open systems semantics. *Artificial Intelligence*, 47:81–107, 1991.
6. S. Ghosh and T. S. Lee. Principles of modeling complex processes. In *Modeling and asynchronous distributed simulation: Analyzing complex systems*, Microelectronic systems, pages 19–30. IEEE Press, 2000.

7. R. Gustavsson and M. Fredriksson. Coordination and control of computational ecosystems: A vision of the future. In A. Omicini, F. Zambonelli, M. Klusch, and R. Tolksdorf, editors, *Coordination of Internet agents: Models, technologies, and applications*, pages 443–469. Springer, 2001.
8. R. Gustavsson, M. Fredriksson, and C. Rindebäck. Computational ecosystems in home health care. In C. Dellarocas and R. Conte, editors, *Fourth international conference on autonomous agents, Workshop proceedings on Norms and institutions in multiagent systems*, pages 86–100, 2000.
9. M.N. Huhns and L.M. Stephens. Multiagent systems and societies of agents. In G. Weiss, editor, *Multiagent systems: A modern approach to distributed artificial intelligence*, pages 79–120. The MIT Press, 1999.
10. N.R. Jennings. An agent-based approach for building complex software systems. *Communications of the ACM*, 44(4):35–41, 2001.
11. T. Malsch. Naming the unnamable: Socionics or the sociological turn of/to distributed artificial intelligence. In N. Jennings and K. Sycara, editors, *Autonomous agents and multi-agent systems*, volume 4, pages 155–186. Kluwer Academic Publishers, September 2001.
12. G. Nicolis and I. Prigogine. Complexity in nature. In *Exploring complexity: An introduction*, pages 5–45. W. H. Freeman and Co., 1989.
13. K.R. Popper. Further remarks on reduction, 1981. In *The open universe: An argument for indeterminism*, pages 163–174. Routledge, 1992.
14. G. Schreiber, H. Akkermans, et al. Knowledge engineering basics. In *Knowledge engineering and management: The CommonKADS methodology*, pages 13–24. MIT Press, 1999.
15. D. Tennenhouse. Proactive computing. *Communications of the ACM*, 43(5):43–50, 2000.
16. L. von Bertalanffy. The meaning of general system theory. In *General system theory: Foundations, development, applications*, pages 30–53. George Braziller, 1988.
17. M. Wooldridge. Semantic issues in the verification of agent communication languages. In N. Jennings, K. Sycara, and M. Georgeff, editors, *Autonomous agents and multi-agent systems*, volume 3, pages 9–31. Kluwer Academic Publishers, March 2000.

A Distributed Approach to Design Open Multi-agent Systems

Laurent Vercouter

Dèpartement SMA, Centre SIMMO
Ecole Nationale Supèrieure des Mines de Saint-Etienne
158 Cours Fauriel, F-42023 Saint-Etienne, France
Laurent.Vercouter@emse.fr

Abstract. An open multi-agent system (MAS) is a MAS where agents may be added, be removed or evolve (e.g. modify their abilities). Usually, the openness of a MAS is controlled by a middle-agent. In this paper, we tackle the problem of openness from a general point of view (considering the middle-agent approach as one of the possible solutions). Then, we propose an original approach based on the principle of distribution which requires a collective activity of the agents to manage openness. Finally, we compare these two approaches to determine the advantages and disadvantages of each.

1 Introduction

Multi-agent systems (MAS) are often presented as a powerful paradigm for studying and designing evolutive distributed systems. In this paper, we will focus on a specific kind of evolutive MAS which are open multi-agent systems. An open system should support the addition or the removal of some functions after its design (and generally, during its execution). In a MAS, the global functions of the system are carried out by one or several agents. Thus, an open MAS should support the addition of new agents, the removal of agents and the internal evolution of agents (e.g. modification of their abilities).

Usually, the openness of a MAS is ensured by an entity, called a middle-agent. This approach, first established by Y. Labrou and T. Finin [10], is easy to implement, and is used by most of the existing works on open MAS. Research in this area is mainly concern in designing brokering architectures [1,8], brokering protocols and communication languages [10,6] or formalisms to represent agent capabilities [7,14,16,20]. However, the use of a middle-agent to manage the system openness has some limits and is not suitable in some contexts. The goal of our work is to consider the openness from a general point of view in order to target the main problems that should be solved during the design of an open MAS. Then, we will be able to suggest new solutions in order to tackle the problem of openness.

A. Omicini, P. Petta, and R. Tolksdorf (Eds.): ESAW 2001, LNAI 2203, pp. 25–38, 2001.

2 Features of an Open MAS

The functions available in a MAS may be executed by an agent or by the way of a cooperation involving several agents. In the context of an open MAS, we focus on two points: the composition of a MAS (its number of agents and the distribution of capabilities among them) and the possibilities of cooperations between agents. An open MAS should support and deal with three possible modifications of its composition:

- **Addition** of an agent to the MAS (in order to provide new capabilities for example);
- **Removal** of an agent from the MAS (if its capabilities are obsolete for example);
- **Evolution** of an agent (the agent gains new capabilities and loses some of its old capabilities).

These modifications are "physical" alterations of the MAS which have a great influence on the possibility of cooperations between agents. Indeed, the addition of an agent to a MAS creates new possibilities of cooperation between this agent and some other agents of the system, whereas the removal of an agent causes some of these possibilities to become obsolete. When an agent evolves, its possibilities of cooperation also change because it has a different set of capabilities and different needs.

Such modifications break the stability of the system since some possibilities of cooperations between agents are no longer valid and that new possibilities exist but are unknown by the agents. Therefore, the management of an open MAS mainly consists in detecting the possibility of cooperations between agents. Whichever entity is responsible for this detection, it has to know the characteristics of the agents by the way of *agent representations.* Each representation should contain a set of descriptors which may be compared to check if two descriptors are *complementary.* Then, a possibility of cooperation between two agents ω_i and ω_j may be detected if there exists a complementarity between one descriptor of ω_i and one descriptor of ω_j.

Openness management is closely related to the problem of complementarities detection (which can be considered as a part of the connection problem [4]). Checking the success of the addition, removal or evolution of an agent requires the introduction of a third notion: the *integration* state of an agent. An agent would be considered as *integrated* if and only if every complementarity between each of its descriptors and each of the descriptors of all other agents of the MAS have been detected. To give a formal definition of the complementarity between two descriptors, we need to define the concepts handled in an agent representation (what is done in section 3). In section 4, we study different approaches of openness management, each of them having its own evaluation of the integration state of an agent, and we compare them in section 5. Section 6 presents our directions for future work.

3 Agent Modeling in an Open MAS

Several formalisms for agent representations have been proposed in related work. Some of them focus on the mental state [12,13] on the decision process [2,9], or on the capacities [7,16,17] of the agents. For our purpose, the used formalism should be suitable to calculate complementarities between two agents. According to G. Wickler [20], agent representations should be based on their capacities in order to be used for brokering. Thus, an agent which has a specific capability that can address a given problem can be found by searching the set of representations.

K. Decker *et al.* [5] distinguished two types of information that should be contained in an agent representation: capabilities and preferences. Capabilities describe the kind of services that an agent can provide and preferences are the services that an agent may require. In our representation formalism, we define these two types of information. This formalism is quite close to the one defined by J. Sichman [15,14] who suggested a model for social reasoning based on the dependence theory [3]. As dependence detection and complementarities are similar problems, Sichman's formalism only needs a few adjustments in order to be suited to openness management.

We define an *agent description* (AD_i, such that i is the id of the represented agent) as being composed of two parts: (i) a *functional description* (FD_i) to represent its goals and capacities, (ii) a *cooperative description* (CD_i) to represent its requirements (corresponding to the concept of preferences mentioned by K. Decker *et al.*). Besides of the descriptors contained in these two parts, an agent should also have some meta-knowledge about these descriptors which define their semantics. We consider that this meta-knowledge are classes and that a descriptor is an instance of one of these classes.

3.1 Functional Description

The functional description of an agent ω_i is a quadruplet $FD_i = (G_i, P_i, A_i, R_i)$ such that:

- G_i is the set of agent i's goals;
- P_i is the set of agent i's known plans;
- A_i is the set of agent i's available actions;
- R_i is the set of agent i's available resources.

Each of these descriptors refers to a class of descriptors. Thus, an agent's meta-knowledge contains goal classes, plan classes, action classes and resource classes which are structured in trees in order to express different levels of generality (more details on the structure of these classes may be found in [19]). The assumption is made that the agents' meta-knowledge comes from a common ontology and that they are compatible. Each agent's meta-knowledge is a part of this ontology.

Some functional descriptors represent goals or capabilities which may need the use of other capabilities. For example, a plan achieves a goal by the accomplishment of a few actions or sub-plans, and an action may use and produce some resources. These relations can be used to automatically deduce an agent's requirements which are represented in a cooperative description.

3.2 Cooperative Description

The cooperative description of an agent ω_i is a triplet $CD_i = (PR_i, TR_i, RR_i)$ such that:

- PR_i is a set of plan requirements. If the agent i has a goal of a class g^*, it requires to know which of its acquaintances knows a plan that can achieve a goal of the class g^*. Thus, the agent ω_i has a plan requirement $pr \in PR_i$ such that pr designates the goal class g^*. Each plan requirement is described by the goal class that has to be achieved;
- TR_i is a set of task requirements. We use the word "task" to refer to a plan or an action (without knowing if the descriptor is a plan or an action). A task requirement contains only the task class that is required. An agent ω_i has a task requirement $tr \in TR_i$ about the task class t^* if and only if it knows a plan $p \in P_i$ in which t^* may be accomplished;
- RR_i is a set of resource requirements. A resource requirement is described by a resource class. An agent ω_i has a resource requirement $rr \in RR_i$ about the resource class r^* if and only if it may accomplish an action $a \in A_i$ which may use a resource of class r^*.

We suggest here to deduce an agent's requirements from its capabilities but it is also possible to determine them by another way (some other algorithms or at the creation of the agent).

3.3 Complementarity of Descriptors

We can now define a predicate to evaluate the complementarity between an capability c and a requirement r:

$$\boxed{compl(c, r) \equiv class(c) \leqslant class(r)}$$

$class(c)$ (or $class(r)$) is the descriptor class of the capability c (or r) and the symbol $\leqslant$ checks if the first class is the same or a specialization of the second class. For example, if an agent ω_i can use a resource which refers to a resource class $ColorPrinter$ and an agent ω_j has a resource requirement about the resource class $Printer$, these two descriptors would be complementary if and only if the resource class $ColorPrinter$ is a specialization of the resource class $Printer$.

Agent representations are essential in an open MAS to detect any changes in the possibilities of cooperation between agents and to maintain their integration state. We consider that an agent has a true and complete agent description about itself and that it only knows a partial agent description for a few other agents. Then, the integration state of an agent depends on the content of these descriptions about others. In the next section, we present how this integration state may be maintained in an open MAS and study two main approaches.

4 Openness Management

Cooperation possibilities are deduced from a comparison of agent descriptions in order to find complementarities. This implies that there should exist one or more agents which know several agent descriptions. Furthermore, modifications of the composition of an open MAS may render some agent descriptions invalid or incomplete. Then, some mechanisms to update the agent descriptions (wherever they are) must be defined in order to manage the openness of the MAS. The figure 1 illustrates these problems and introduces the distinction criteria between the two main approaches that we study.

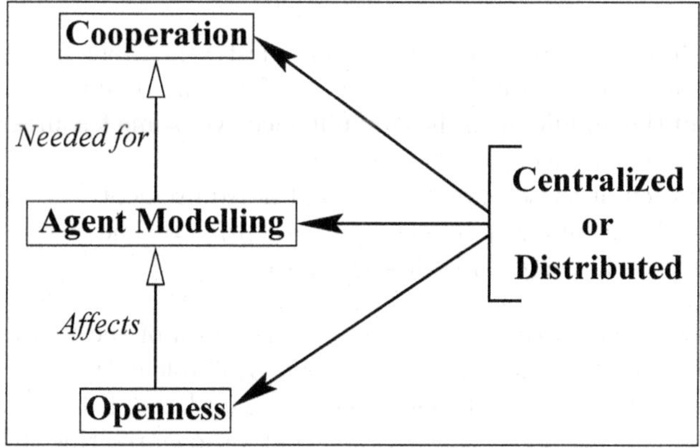

Fig. 1. Openness Management Approaches.

This distinction relies on the number and the roles of agents which are in charge of openness, agent modeling and cooperation detection.

4.1 Centralized Approach

In this approach, a specific agent is included in the MAS to deal with openness, to maintain agent representations and (sometimes) to find cooperation possibilities. Most of the existing works on open MAS follow this approach, defining such an agent, usually called a middle-agent. This solution is very popular because it is easy to implement and openness can be achieved by simply creating a middle-agent. The other agents of the MAS do not have to be modified to support openness except that they must know how they should interact with the middle-agent.

Several kinds of middle-agent (see table 1) are distinguished by K. Decker *et al.* [5]. The role of each middle-agent depends on its knowledge and on the knowledge of the agents which have some capabilities (the providers) and of the

Table 1. Different Roles for a Middle-Agent (see [5] for More Details).

requirements known by	*Capabilities known by* **provider**	**provider + middle-agt**	**provider + middle-agt + requester**
requester	(broadcaster)	"front agent"	matchmaker / yellow-pages
requester + middle-agt	anonymizer	broker	recommender
requester + middle-agt + provider	blackboard	introducer/ bodyguard	arbitrator

agents which have some requirements (the requesters). An agent may be both a provider for some capabilities and a requester for some requirements.

The type of the middle-agent has an influence over some features of the MAS. For example, if capabilities are not known by the requesters, the system may ensure the privacy of these informations and if requirements are not known by the providers, the agents' requirements may be private. The characteristics of a centralized approach, using a middle-agent, are:

- **Openness** (*centralized*): each time a modification of the MAS composition occurs, the set of abilities and requirements available in the system change. If the knowledge of the middle-agent should be updated, it has to be informed about this change. Then the middle-agent propagates this information to providers or requesters according to the amount of knowledge they should have.
- **Agent modeling** (*centralized*): in most of the case, the middle-agent knows every agent in the system and maintains a description for each of them. It keeps this knowledge available for the agents of the system and can communicate part of it to providers and requesters.
- **Cooperation detection** (*centralized or distributed*): Cooperation detection needs that an agent knows both the capabilities and requirements of the agents of the MAS in order to compare them and find complementarities. If providers and requesters has some knowledge about capabilities and requirements, they are able to detect cooperation possibilities. In all the other cases, the middle-agent is needed as an intermediary for cooperation detection. The middle-agent may only give some informations about other agents (e.g. enumerate all the providers which have a given capability) or it may establish a cooperation on behalf of another agent (e.g. find all the providers matching a given requirement, propose to some of them to cooperate and inform the requester of the chosen partner).

To give a definition of the integration state of agents we need to add an index to the notations previously introduced in order to represent which agent has the specified knowledge. For example, we note AD_j^i the agent description known by an agent ω_i to represent an agent ω_j. To designate the middle-agent, we use the index ma. Then the integration state of the agents depends on the kind of middle-agent that is used. If a broker is used, the integration state of an agent ω_i is:

$$\boxed{integrated(\omega_i) \equiv (AD_i^i = AD_i^{ma})}$$

This definition ensures that an agent ω_i is integrated if and only if the broker knows its agent description and that it is up to date according to the the capabilities and requirements of ω_i. To deal with the removal of an agent, we assume that the agent description is specific to the concerned MAS and that it should indicate if the agent has left the MAS (for example, by the use of a flag or by setting the agent description to null, $AD_i^i = \emptyset$).

If providers should know the requirements of other agents the following formula has to be added to the integration state:

$$\boxed{\forall \omega_j \in \Omega, \forall a \in FD_i^i, \forall r \in CD_j^j, compl(a,r) \supset r \in CD_j^i}$$

Such that Ω is the set of all the agents of the MAS. If requesters should know the capabilities of other agents the following formula has to be added to the integration state:

$$\boxed{\forall \omega_j \in \Omega, \forall r \in CD_i^i, \forall a \in FD_j^j, compl(a,r) \supset a \in FD_j^i}$$

The use of a middle-agent has first been suggested by Y. Labrou and T. Finin [10] and has been specified by the FIPA [6]. The agent platform of the FIPA is open since it provides the middle-agent services by the way of two agents: the *agent management system* (which provides a naming service) and the *directory facilitor* (which provides yellow pages services).

The centralization of the agent representations and the openness management seems suitable and efficient in order to maintain the integration state of the agents. Nevertheless, the middle-agent is a critical entity which is essential to openness and cooperation in the MAS. A centralized approach has two drawbacks [11]: (i) the MAS is vulnerable to middle-agent failure (cooperations would be impossible), (ii) since cooperation between agents needs the middle-agent, it may be overloaded by messages and knowledge.

4.2 Distributed Approach

In their classification, K. Decker *et al.* [5] (see section 4.1) mentioned that the lack of a middle-agent implies that the agents have to communicate by broadcast messages. We suggested [18] a new approach to deal with openness, without any middle-agent, and where agents do not communicate by broadcast messages. Each agent has its own representation of other agents, can detect its possibilities of cooperation and is involved in openness management. A distributed approach is characterized by:

- **Openness** (*distributed*): A modification of the MAS composition implies that the agent representations of some agents must be updated. The openness management requires a collective activity of the agents to find out which are these agents in order to inform them of this modification. To ensure that this collective activity leads to a successful openness management, we make the assumption that the agents are cooperative and trustful in their interactions.
- **Agent modeling** (*distributed*): Each agent knows only a few other agents and only a few descriptors about them. To select which agents and which descriptors should be known, we defined the notion of *relevance* of an agent description. An agent ω_i has a relevant representation of an agent ω_j if every descriptor that ω_i knows about ω_j is complementary to one of ω_i's own descriptors. Then, a relevant representation of another agent does not contain any descriptor which is useless for cooperation detection. To compare the notions of relevance and integration, we can say that the relevance requires that only useful informations about other agents are known whereas integration ensures that all useful informations about other agents are known.
- **Cooperation** (*distributed*): As every agent has a relevant representation of other agents, each of them knows its own possibilities of cooperation with others. Therefore, they can locally select their partners and directly propose them to cooperate.

The integration state of an agent depends on the capabilities and requirements of the other agents of the MAS. This state corresponds to the following formula:

$$\begin{array}{l} integrated(\omega_i) \equiv \\ \quad \forall \omega_j \in \Omega, \forall r \in CD_i^i, \forall a \in FD_j^j \\ \qquad compl(a,r) \supset a \in FD_j^i \\ \quad \wedge \\ \quad \forall \omega_j \in \Omega, \forall a \in FD_i^i, \forall r \in CD_j^j \\ \qquad compl(a,r) \supset r \in CD_j^i \end{array}$$

One agent is integrated if and only if it knows (i) all agents having a descriptor complementary to one of its own descriptors and (ii) if it knows, about each of these agents, all these complementary descriptors. In a distributed approach, the agent targeted by this modification (the one which is added, removed or which evolves) has the responsibility of updating its own representation of others and of notifying the modification to the concerned agents. According to the kind of the modification, the agent accomplishes such updates with the help of other agents in different ways:

- **Addition**: it should find the agents with descriptors complementary to some of its own descriptors;
- **Removal**: the agent should only notify its acquaintances that it is leaving the MAS (only its acquaintances know the agent to remove);

- **Evolution**: it is a combination of the operations needed for the two previous modifications. For every descriptor that disappeared, the agent should notify it to the other agents which have a complementary descriptor, and for its new descriptors, it has to find the relevant agents.

Presentation Protocol.

To reach integration, the added or evolved agent initiates a presentation protocol (shown in figure 2) with another agent. This first other agent may be randomly chosen.

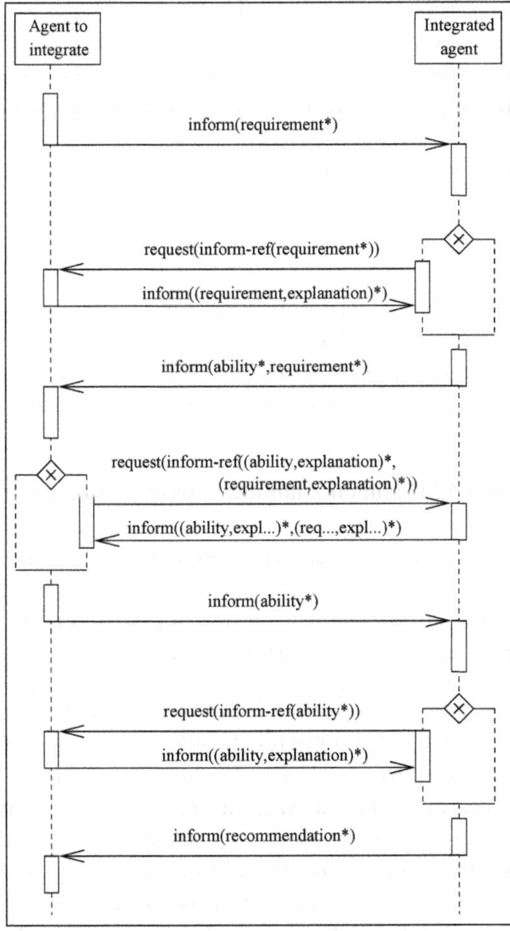

Fig. 2. The Presentation Protocol.

This protocol is basically composed of two parts:

- **Presentation**: The agent to integrate (ω_i) informs the other (ω_j) of its requirements. If some of these requirements refers to some meta-knowledge that ω_j does not know, this one may ask and receive explanations about the semantics of some requirements (their position in the tree of classes). Then ω_j selects its capabilities which are complementary to one of the communicated requirements, and all its own requirements to send these informations to ω_i. Here again, explanations may be requested. At last, ω_i selects its capabilities complementaries to one of ω_j's requirements and send them (with an explanation if required). At the end of the presentation, each agent has a relevant representation of the other.
- **Recommendation**: In order to be integrated, the agent ω_i may need to know some other agents than the one involved in the presentation protocol. Thus, ω_j recommends to ω_i to start a presentation protocol with some of its own acquaintances. Then, ω_i selects an agent ω_k from the list of recommended agents, initiates a new presentation protocol with it, acquires a relevant representation of ω_k and collects a new list of recommended agents.

Recommendations.

The recommendation step is very important for a distributed approach. We do not want that the agent ω_i introduces itself to every agent of the MAS, so it must be able to estimate its integration state on the basis of the recommendations of other agents. We defined several kinds of recommendations (to get more details see [18]):

- **Direct**: ω_j knows all the agents having a descriptor which is complementary to a descriptor of ω_i;
- **Indirect**: ω_j knows at least one other agent which can give a direct recommendation to ω_i;
- **Partial**: ω_j knows some agents (but not all of them) having a descriptor which is complementary to a descriptor of ω_i;

Direct recommendations are essential in order to allow an agent to evaluate its own integration state. If ω_j knows all the agents having a descriptor which is complementary to a descriptor d (it is the case if ω_j is integrated and if $d \in AD_j^j$), ω_j sends a direct recommendation to ω_i about all these agents. According to the definition of the integration state, ω_i *considers itself as integrated if* ω_i *has received a direct recommendation about each of its own descriptors.*

The goal of an indirect recommendation is to identify an agent which can provide a direct recommendation for a given descriptor. Partial recommendations are used for the descriptors which can not be subject to a direct or indirect recommendation, but no deduction is possible for the integration state. If some of ω_j's acquaintances are not involved in a recommendation, ω_j sends their address to ω_i in a plain message (no recommendation).

Recommendation Use.

The agent ω_i uses the received recommendation to initiate new presentation protocols. For its next presentation, it will choose (in a decreasing priority):

- The agent which has the most indirect recommendations;
- The agent which has the most partial recommendations;
- The agent which has the most direct recommendations;
- any other agent.

This sequence of presentations stops in two cases: (i) ω_i consider itself as integrated, or (ii) the list of agent that ω_i knows and which have not been contacted for a presentation is empty. In this second case, the agent ω_i may consider that it has contacted all the agents of the MAS (which may be true if some assumptions, explained below, are made) and then believe that it is integrated. At the end of the presentations, ω_i should communicate with the agents which are concerned by a direct recommendation (if it has not been done in a presentation) to notify them its existence and their complementarities.

To ensure that an agent will be integrated, a few assumptions and constraints should be adopted. The presentations and recommendations are efficient if:

- an agent which is not integrated can only formulate a direct recommendation for a descriptor if it has received a direct recommendation for this descriptor;
- a presentation protocol is a critical section (an agent cannot be involved in two presentation protocols at the same time).

These constraints allow simultaneous modifications of the MAS (i.e. more than one agent may be trying to reach integration at the same time). In the worst case, we mentioned that an agent may have to contact all other agents of the MAS. This is possible if:

- for any two agents of the system, there must exist an acquaintance path between them (i.e. if we consider a graph where the nodes represent the agents and one edge represents the fact that the "linked" agents have a representation of each other, the graph has to be connected).

To keep this assumption true, the removal of an agent should be done carefully. When an agent wants to leave the MAS, it should check if its removal will affect the graph connectivity. If it is true, this agent should notify this to all its acquaintances in order to create "fake" acquaintance links in order to preserve the connectivity.

Our proposition has been implemented as a partial social agent model. The knowledge and functions needed by an agent in a distributed open MAS are grouped in a *welcoming facet.* An agent with a welcoming facet (we call such an agent a *welcoming agent*) can detect complementarities between descriptors, initiate or participate in a presentation protocol, and formulate recommendations. Then, it is possible to build an open and distributed MAS if it is composed of welcoming agents.

Compared to the centralized approach, the use of welcoming agents is interesting because no agent is essential for openness or cooperation. Therefore, such an open system is robust because an agent failure does not prevent the other agents to cooperate and the system remains open. The cost of this robustness is that openness management needs more communications.

5 Comparison of the Approaches

These two approaches of openness management can be compared by few criterias:

- the **number of agents to communicate with** to deal with a modification of the composition of the MAS (**openness**) or to find a cooperation partner (**cooperation**);
- the **relevance of the representation of other agents**: a representation that is not relevant implies a cost to select relevant informations while seeking a cooperation partner and contains useless informations;
- the **robustness of the system**: the MAS is robust if it can support agent failures.

The features of each approach of openness management are summarized in the table 2.

Table 2. Comparison of the Centralized and Distributed Approaches.

	Communications		Relevance	Robustness
	Openness	Cooperation		
Centralized	1	2	No	No
Distributed	$\leqslant \Omega$	1	Yes	Yes

In a centralized approach, the cost in communication is very low: the modification of the MAS composition requires only the sending of a message to the middle-agent. Following our distributed approach, an agent may have to contact all the agents of the MAS. Thus, the number of agents to contact is bounded by the total number of agents (Ω). Seeking a partner to cooperate is a bit more costly if a middle-agent is used but this low difference may be important for two reasons: (i) in most of the systems, there is much more cooperations between agents than additions, removals or evolutions of agents, and (ii) the middle-agent is always involved in these communications and may be overloaded by the request of other agents.

The main advantages of our proposition are the relevance of the representation of other agents and the global robustness of the MAS. The distributed approach may be appropriate in contexts where the agents' knowledge are limited in size (useless informations about others may be forgotten) and where agent failures may occur. For example, it seems suitable to conceive open MAS in domains like robotics or nomadic computing.

We are now working on an experimental comparisons of the centralized and distributed approach. These tests would allow us to determine empirically the cost in communications of our proposition, and to increase the precision of the comparison of the approaches.

6 Future Work

In this paper, we target the main concepts (complementarity, integration, relevance, recommendation) that are taken into account during the openness management. We were interested in their application and definition according to a centralized or a distributed approach. However, these two categories of approaches are very general and there exists some open MAS which follow hybrid approaches.

Our final goal is to define a generic framework for openness, based on the concepts presented above. Thus, our future work will be to describe these hybrid approaches and to continue our generalization of the openness to build a framework independent from the approach chosen or from the agent representation structure. Moreover, we are interested in the definition of different kinds of agent integration. For example a *partial integration*, where an agent only knows a sub-set of its possible partners for cooperation, may be appropriate in some contexts.

Our work will also be implemented on a concrete application for the E-Alliance[1] project (funded by the Région Rhône-Alpes Thematic Program). The aim of this project is to build a software infrastructure for the management of an alliance of several printshops. This infrastructure should allow information exchange, cooperations and negotiations between printshops and should be open to new printshops or to other partners (for the delivery of the print jobs for example).

References

1. R.J. Bayardo, W. Bohrer, R. Brice, A. Cichocki, J. Fowler, A. Helal, V. Kashyap, T. Ksiezyk, G. Martin, M. Nodine, M. Rashid, M. Rusinkiewicz, R. Shea, C. Unnikrishnan, A. Unruh, and D. Woelk. Infosleuth: Agent-based semantic integration of information in open and dynamic environments. In Michael Huhns and Munindar Singh, editors, *Readings in Agents*, pages 205–216. Morgan Kaufmann, 1998.
2. David Carmel and Shaul Markovitch. How to explore your opponent's strategy (almost) optimally. In *Proceedings of the Third International Conference on Multi-Agent Systems*, pages 64–71, Paris, France, July 1998. IEEE Computer Society.
3. Cristiano Castelfranchi, Maria Miceli, and Amedeo Cesta. Dependence relations among autonomous agents. In Eric Werner and Yves Demazeau, editors, *Decentralized A. I. 3*, pages 215–227. Elsevier Science Publishers, 1992.
4. Randall Davis and Reid G. Smith. Negociation as a metaphor for distributed problem solving. *Artificial Intelligence*, 20(1):63–109, 1983.

[1] `http://www.xrce.xerox.com/ealliance/`.

5. Keith Decker, Katia Sycara, and Mike Williamson. Middle-agents for the internet. In *Proceedings of the 15th International Joint Conference on Artificial Intelligence*, pages 578–583, Nagoya, Japan, August 1997. Morgan Kaufmann.
6. Foundation for Intelligent Physical Agents. Agent management specification. Technical Report XC00023G, Foundation for Intelligent Physical Agents, Geneva, Switzerland, August 2000.
7. Les Gasser, Carl Braganza, and Nava Herman. Implementing distributed ai systems using mace. In *Proceedings of the 3rd IEEE Conference on Artificial Intelligence Applications*, pages 315–320, Orlando, February 1987.
8. Marie-Pierre Gleizes and Pierre Glize. Abrose: Des systèmes multi-agents pour le courtage adaptatif. In *Journées Francophones d'Intelligence Artificielle Distribuée et de Systèmes Multi-Agents*, pages 117–132, Saint-Jean-La-Vêtre, Loire, October 2000. Hermès. (in french).
9. Piotr J. Gmytrasiewicz and Edmund H. Durfee. A rigorous, operational formalization of recursive modeling. In *Proceedings of the first International Conference on Multi-Agent Systems*, pages 125–132, San Francisco, California, June 1995. AAAI Press/MIT Press.
10. Y. Labrou and T. Finin. A semantics approach for kqml - A general purpose communication language for software agents. In *Third International Conference on Information and Knowledge Management*, 1994.
11. Vladimir Marik, Michal Pechoucek, and Olga Stepankova. Social knowledge in multi-agent systems. In M. Luck, V. Marik, O. Stepankova, and R. Trappl, editors, *Multi-Agent Systems and Applications, ACAI 2001*, pages 211–245, Prague, Czech Republic, July 2001. Springer.
12. Anand S. Rao and Michael P. Georgeff. Modeling rational agents within a bdi architecture. In J. Allen, R. Fikes, and E. Sandwall, editors, *Proceedings of the Int. Conf. on Principles of Knowledge Representation and Reasoning, KR-91*, pages 473–484, San Mateao, 1991. Morgan Kaufmann.
13. Yoav Shoham. Agent-oriented programming. Technical report, Stanford University, 1990.
14. Jaime S. Sichman and Yves Demazeau. Exploiting social reasoning to deal with agency level inconsistency. In V. Lesser, editor, *Proccedings of the first International Conference on Multi-Agent Systems*, pages 352–359, San Francisco, California, June 1995.
15. Jaime Simao Sichman. *Du Raisonnement Social chez les Agents.* PhD thesis, Institut National Polytechnique de Grenoble, September 1995. (in french).
16. Peter Stone, Patrick Riley, and Manuela Veloso. Defining and using ideal teammate and opponent agent models. In AAAI Press/MIT Press, editor, *Proceedings of the Seventeenth National Conference on Artificial Intelligence (AAAI-2000)*, pages 1040–1045, Austin, Texas, August 2000.
17. Milind Tambe. Towards flexible teamwork. In *Journal of Artificial Intelligence Research*, volume 7, pages 83–124. Morgan Kaufmann, 1997.
18. Laurent Vercouter. *Conception et mise en oeuvre de systèmes multi-agents ouverts et distribués.* PhD thesis, ENSM.SE, UJM, December 2000. (in french).
19. Laurent Vercouter, Philippe Beaune, and Claudette Sayettat. Towards open distributed information systems by the way of a multi-agent conception framework. In Yves Lespérance, Gerd Wagner, and Eric Yu, editors, *Agent-Oriented Information Systems, Seventeenth National Conference on Artificial Intelligence*, Austin, Texas, July 2000.
20. Gerhard Wickler. *Using Expressive and Flexible Agent Representations to Reason about Capabilities for Intelligent Agent Cooperation.* PhD thesis, University of Edinburgh, 1999.

Engineering Infrastructures for Mobile Organizations

Giacomo Cabri, Letizia Leonardi, Marco Mamei, and Franco Zambonelli

Dipartimento di Scienze dell'Ingegneria – Università di Modena e Reggio Emilia
Via Vignolese 905 – 41100 Modena – Italy
{giacomo.cabri,letizia.leonardi,mamei.marco}@unimo.it,
franco.zambonelli@unimo.it

Abstract. Mobile application components can be effectively and uniformly modeled in terms of autonomous agents moving across different contexts during execution. In this paper, we introduce a conceptual framework based on the definition of programmable organizational contexts, which can promote an engineered approach to application design and that, if is supported by a proper programmable coordination infrastructure, can make applications more modular and easy to maintain. On this base, the paper analyses several issues related to the implementation of programmable coordination infrastructures for mobility. In addition, the paper introduces a preliminary proposal for the modeling of programmable coordination infrastructures in terms of a general-purpose event-based infrastructure. Finally, the paper sketches open issues and promising research directions.

1 Introduction

The wideness and the openness of the Internet scenario makes it suitable to design applications in term of components that are aware of the distributed nature of their own environment, and that are able to move trough this environment.

Mobility may appear in different flavors in today's computing environments:

- *Virtual mobility*. Components *virtually move* across Internet resources [7] by accessing resources and interacting in a network-aware fashion. They "navigate" the Internet by explicitly locating services and resources.
- *Actual mobility*. This refers to the components' capability of moving across Internet sites while executing by dynamically and autonomously transferring their code, data and state, toward the resources they need to access [19].
- *Physical mobility*. Mobile devices – such as palmtop, cellular phones, etc. – accessing the Internet and/or interacting with each other in the context of mobile ad-hoc network (MANETs) will be more and more present in application scenarios, and they will have to be properly handled and modeled [16].

A. Omicini, P. Petta, and R. Tolksdorf (Eds.): ESAW 2001, LNAI 2203, pp. 39-56, 2001.

To limit complexity of application design and development, suitable models and infrastructures are needed to handle the different types of mobility in a natural and uniform way.

A promising approach to uniformly deal with all the above kinds of mobility is to model application components, as well as physical mobile devices and computer-based systems, in terms of *autonomous mobile agents*. In general terms, agents are autonomous entities capable both of reacting to changes in the environment and of executing in a proactive way, and often typically implemented by active objects integrating event-handling and exception-handling capabilities [10]. These characteristics imply that:

- Agents are provided local control over their activities, thus enabling dealing in a natural and decentralized way with the openness, dynamism, and unpredictability of the Internet and, more generally, of dynamic and decentralized networked scenarios;
- Agents are explicitly designed to be situated in an environment. This naturally invites thinking mobile application components in terms of agents that situate in different environments during their lives.
- Physical mobility, unlike virtual and actual ones, cannot usually be controlled at the application level, and a useful way of modeling it is in terms of an additional dimension of autonomy of application components. In other words, even if no agent-technology is exploited inside a mobile device, its observable behavior w.r.t. the fixed infrastructure or to other devices is actually those of an autonomous component, i.e., of a mobile agent, and will have to be modeled accordingly.

In the following of this paper, we detail how the above perspectives on mobility can drive the definition of a suitable organizational framework for the design of applications (Section 2). Because the organizational framework needs to be supported, at the middleware level, by a proper programmable coordination infrastructure for application development and execution, we analyze the main issues arising in the definition of such an infrastructure (Section 3). In addition, we discuss our preliminary attempts at modeling programmable coordination infrastructures for mobility via the definition of a general-purpose event-based kernel (Section 4). Finally, we sketch what are, in our opinion, the key research directions in the area of middleware and coordination infrastructure for agent-based mobile computing (Section 5).

2 The Conceptual Framework

2.1 Local Interaction Context

Handling mobility requires facing different issues at different levels, also depending on the type of mobility. However, as far as the high-level issues related to the modeling, design and development of complex multi-component applications are concerned, handling mobility is basically a problem of managing the coordination activities of application agents. These may include accessing the local resources of an environment and communicating and synchronizing with executing agents, whether belonging to the same application or foreign Internet agents.

In our approach, we model mobility of agents across the Internet as movements across *local interaction contexts*. A local interaction context defines the agents' perceivable world, which changes depending on the agent position, and which represents the logical place in which agents' coordination activities occur. What is the interaction model to be actually exploited for modeling interaction in a context is not of primary influence. In fact, for the fact of moving across the Internet, agents access different data and services, and interact with different agents, depending on their current interaction context, independently of the model via which interactions occur. Interactions may occur via message passing and ACLs [6], via meetings [19], or via shared dataspaces [1, 8]. What really matters from the software engineering perspective is the locality model enforced, which reflects – at the level of application modeling – a notion of context intrinsic in mobility.

We disregard the modeling of the *execution contexts* intended as the places in which the agents actually execute. In this way, our model can be defined independently of the type of mobility exhibited by application agents (i.e., virtual, actual, or physical), fully disregarding the specific mobility issues.

2.2 Local Organizations

Movements across local interaction contexts may impact on agents' coordination activities. In fact, in the open Internet scenario, one cannot conceive that agents' coordination activities can be totally free and unregulated. Instead, coordination activities in the Internet are likely to be strictly ruled by proper security and resource control policies, which may be different from site to site, i.e., from a local interaction context to another. Moreover, each local interaction context is likely to adopt peculiar choices for the representation of local resources, and it is likely to host the coordination activities of different agents, and of making available different services to agents. In such a scenario, the local interaction context cannot be simply considered as the place in which coordination activities occur, but it is also an active environment, capable of enacting specific *local coordination laws* to rule and support the agents' coordination activities.

By assuming an organizational (or social) perspective [20, 21], one can consider the local interaction context in terms of an organizational (or social) context. In facts, one must consider that an agent, by entering via a movement an interaction context, enters a foreign organization (society) [23] where specific organizational rules (or social conventions) are likely to be enacted in the form of coordination laws. Thus, for the sake of conceptual simplicity, one must consider the interaction context as the locus in which the organizational laws ruling the activities of the local organization reside [21].

2.3 Application-Specific Organizations

The above is not the full picture. In fact, apart from the previous presented issue. Agents may be part of a cooperative multi-agent application, and move in the Internet to cooperatively achieve, according to specific protocols and patterns, specific application sub-goals. Given that, it is clear that agents' coordination activities within a multi-agent application may not be fully deregulated but, again, may be required to occur accordingly to specific laws that rule the whole application and ensure the proper achievement of the application goal. In other words agents, belonging to a specific application, can logically constitute their own application-specific organizational context.

By assuming if fact the organizational perspective, the context in which an agent executes and interacts is not only the one identified by the local organizational context, but is also the one of its own organization. As that, a local organizational context should not only be thought as the place in which local organizational rules reside, but also as the active context in which application agents may enact their own, application-specific, organizational rules in the form of coordination laws.

2.4 Designing Applications around Programmable Organizational Contexts

The above analysis suggests modeling and designing applications in terms of agents interacting via *active organizational contexts*. Contexts are no longer merely the place in which agent's coordination activities occur, but they become the place where both local and application-specific organizational laws reside and are enacted, via programmability of the behavior of the interaction spaces.

The adoption of such a conceptual framework – that we have defined *context-dependent coordination* [2] – can have a very positive impact on the engineering of mobile agent applications. From the point of view of application designers, the framework naturally invites in designing an application by clearly separating the intra-agent aspects and inter-agent (organizational) ones. The formers define the internal behavior of agents and its observable behavior. The latters define the application-specific organizational laws according to which agents should interact with each other and with external entities for the global application goal to be coherently achieved, and lead to the identification of the coordination laws that agents should spread on the visited interaction contexts. This separation of concerns is likely to reduce the complexity of application design and can make it more modular and easy to be maintained (design-for-change perspective).

Independent of the role of the application designers is the role of site administrators. When new kinds of application agents are going to be deployed on the Internet, the administrator of one site can analyze which local organizational laws that (s)he may find it necessary to locally enforce in terms of coordination laws on the local interaction context. These can be used both to facilitate the execution of the agents on a site and to protect it from improper exploitation of the local interaction context. These site-specific laws will work together with the application-specific organizational laws, to be possibly spread by application agents.

3 Implementing Programmable Coordination Infrastructures

The separation of concerns promoted by context-dependent coordination during analysis and design can be preserved during the development and maintenance phases too if a proper coordination middleware infrastructure is available that somehow reflects the concepts and the abstractions of the context-dependent coordination. In that case, the code of the agents can be clearly separated from the code implementing the coordination laws (whether local organizational laws or application-specific ones). Thus, agents and coordination laws can be coded, changed, and re-used, independently of each other.

A coordination infrastructure for context-dependent coordination must be based on an architecture implementing the abstraction of active organizational contexts, i.e., implementing *programmable coordination media* [5]. By programmable coordination medium, we intend the software system mediating and ruling all coordination activities of application agents within a locality, accordingly to user (or agent-specified) coordination laws, embedded into the medium itself.

There are several issues to be handled in the definition of a programmable coordination middleware infrastructure for the handling of agents' coordination activities. Such issues include:

1. Defining an architecture based on a multiplicity of independent and independently programmable coordination media, each associate to a locality scope, and providing for dynamically binding an agent to a coordination medium accordingly to the agents' movements (virtual, actual or physical);
2. Defining the interaction model to be actually supported by the coordination media, e.g., message-based, or tuple-based, or event-based;
3. Enabling a dynamic programming of the behavior of coordination media both by the local administrators and by application agents' themselves;
4. Identifying a suitable model (and the associated language) for the programming of the coordination media behavior.

In the following, we mainly focus on issues 1 and 3 by distinguishing the case in which a fixed network infrastructure is available from the one in which it is not.

3.1 Fixed Network Infrastructures

When agents execute and interact via the support of a fixed network infrastructure, the most natural choice is to conceive coordination media as allocated on the fixed network architecture.

With regard to the architecture of coordination media, several choices can be adopted.

On the one hand, coordination media can be associated to single Internet nodes, to act both as the place via which to access to the local resources of that node and as "agora" for a set of interacting agents. On the other hand, a set of Internet nodes composing a domain can share a single coordination medium, to be exploited for all the agents interacting in the context of that domain. In this last case, the coordination medium can be associated to a single node of the domain, or it can be implemented in a distributed way across a multiplicity of nodes in the domain. It is also worth noting that nothing prevents from applying the last implementation schema to a geographically distributed domain of nodes, which happens when a single organization spans across the Internet. However, in these cases, implementation problems arise in guaranteeing a correct and consistent execution of programmed behaviors. In fact, the behavior of a coordination medium can rely on the access to a possibly large status of the medium itself. Thus, maintaining the consistency of the state of geographically distributed coordination media can incur in high overhead and made the behavior of the medium itself unreliable. These problems may be affordable when the interaction model defined by a coordination medium and the specific behaviors programmed in it are likely to exploit only loosely the state of the media, as it can be the case of a message-based coordination medium. However, in the case of interaction models in which the status of the medium plays a central role, as it can be the case of tuple spaces, the cost of maintaining the consistency on large status may be simply unaffordable. Thus, in general, we can see a federation of geographically distributed nodes not as a set of nodes sharing a single coordination medium, but rather as a set of nodes each implementing its own independent medium, and sharing only the coordination laws enacted in their media. In other words, geographically distributed nodes have to maintain their independence as coordination media, although they may decide to enact the same rules to represent logically a single organization.

In multi agent application, the specific-application organizational context is enforced by giving application agents the capability to dynamically programming the visited coordination media with application-specific organizational laws. In this case, all agents of an application spread the same coordination laws over all visited sites to share the same application-specific organizational laws (in addition to the locally enforced ones).

Enabling dynamic programmability of coordination/organizational laws is basically an issue of *code mobility*. In general, coordination laws represent a computational behavior to be assumed by the coordination media in response to interaction events. Then, as agents interact with different coordination media allocated on different sites during their lives, and they may be in need of programming such media, coordination laws must be necessarily expressed by using mobile code technology [7]. However, it is interesting to show how the specific type of agent mobility may impact on the specific paradigm of code mobility to be adopted for enabling agent to program the accessed coordination media via application-specific coordination laws.

In the case of a virtual mobile agent (Figure 1a), the agent can interact with remote coordination media without actually transferring itself. However, if the agent wants to impose some specific coordination laws on a remote medium, it has to upload to the remote medium the mobile code implementing the coordination laws. That defines a model of "remote evaluation", in that the code is transferred to a remote site and there execute in response to the agent's coordination activities. A similar approach is followed by the TuCSoN coordination model [13], to program the behavior of remote tuple spaces, and by T Spaces, to insert new communication primitives into tuple spaces [9].

In the case of an actual mobile agent (Figure 1b), the agent transfers across the network its code and state, and carries on also the mobile code needed to implement the application-specific coordination laws. This is the approach adopted by the MARS infrastructure for actual mobile agents [1].

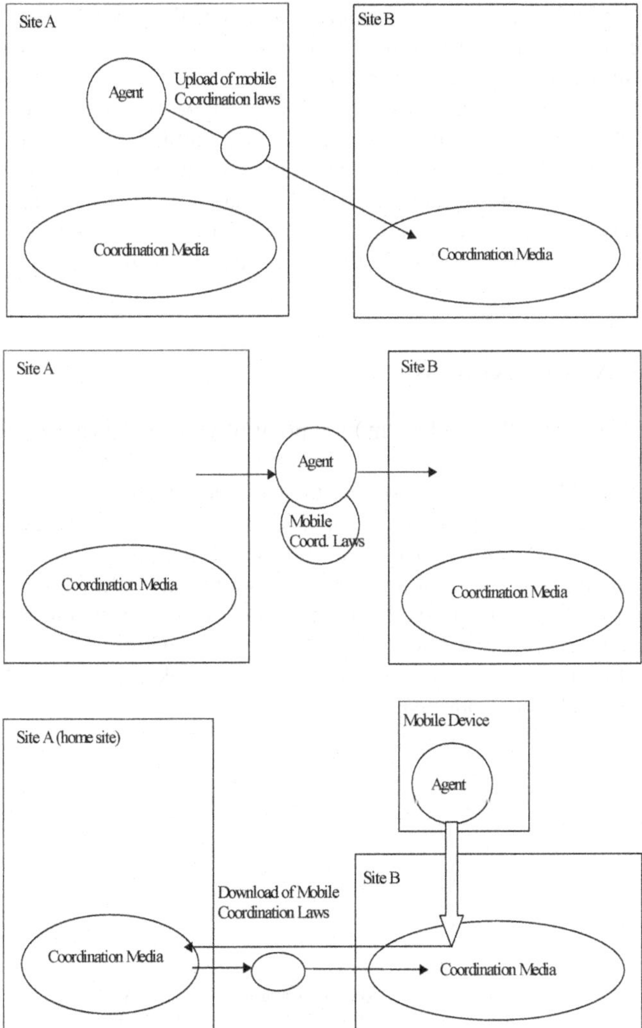

Fig. 1a (top): A virtual mobile agent uploading coordination laws. 1b (middle): An actual mobile agent carrying coordination laws. 1c (bottom): A physical mobile agent causes its current coordination medium to download the coordination laws.

In the case of a physical mobile agent, one can think at having the agent itself (i.e., the physical device it represents) carry the code needed to implement the coordination laws, and install it in the coordination media it connects to during its itinerant life. However, in most of the cases, mobile devices are likely to be resource-limited devices, for which it is often desirable to limit the amount of memory exploited by application as well as the amount of bandwidth. Therefore, a more suitable way for physical mobile agents to program coordination media is to let the coordination media, upon the access of a physical mobile device, download from a specific "home site" the required code (Figure 1c). This defines a model of "code on demand", in which the code is transferred to a coordination medium upon a local request generated from the coordination medium itself. A recent extension of the MARS infrastructure integrates this feature.

In any case, we emphasize that, whatever the type of mobility of application agents and whatever the consequent type of code mobility, they can coexist in a coordination media. In fact, the coordination media, upon notification of the arrival of agents, can discover whether has virtually, actually, or physically arrive, and can act accordingly for properly installing the coordination laws. Thus, all types of mobility can be handled transparently from the point of view of application designers.

3.2 Mobile Ad-Hoc Networks

More challenging from the implementation point of view is the case in which agents have to interact in the absence of any fixed infrastructure, i.e., the case of mobile ad-hoc networks (MANET). We believe that the cases in which a set of mobile devices will have to compulsory interact in the absence of any fixed infrastructure are very limited (e.g., satellite communications will be ubiquitous). Nevertheless, other reasons such as the need of limiting energy consumption or the high costs possibly implied in interacting with a fixed infrastructure may suggest minimizing in any case the exploitation of the services of the fixed infrastructure. The problem, in this case, is that the absence of the fixed infrastructure makes it hard to find a place where to actually allocate coordination media.

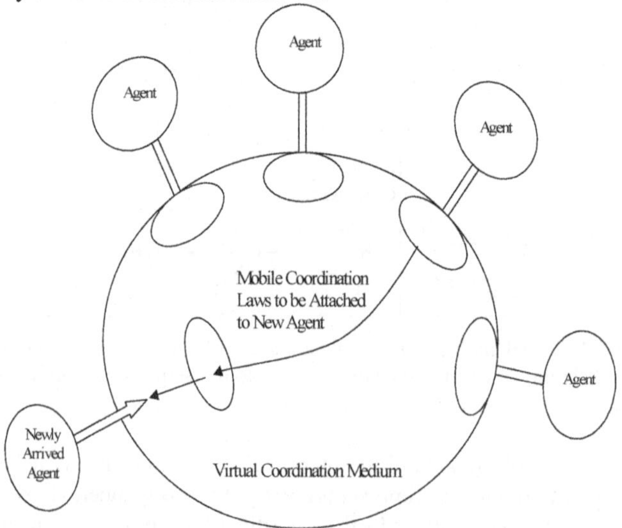

Fig. 2. Programmable Coordination Infrastructure for a MANET.

The basic conceptual framework will not substantially different from the application developers' point of view. However, due to the lacking of a physical network infrastructure, the coordination media via which agents have to interact and in which coordination/organizational laws reside becomes only a *virtual medium*, implemented in a distributed way. The idea is to have a group of agents interacting in the context of a mobile ad-hoc network being somehow "attached" to a software substrate capable of influencing their coordination activities and, thus, of enacting specific coordination laws. When an agent arrives within a MANET, i.e., get connected to a new group and to the corresponding virtual coordination medium (see Figure 2), the code implementing the coordination laws and the software substrate needed to enact them is dynamically uploaded locally to the agent and attached to him to filter all its coordination activities. Such a substrate, again, can exploit mobile code technology to transfer coordination laws from an agent to another one. The substrate, in itself, can be implemented and attached to the agents using different techniques, such as aspect-oriented or reflective programming, or it can be an autonomous software component in its turn, acting as a mediator between the agent and the network. A similar approach has been followed by the LGI model [11] and within the FishMarket project [12]. The way of influencing the behavior of agents by dynamically attaching coordination laws to them strongly resembles the above approach. However, these proposals take into account neither mobility nor the absence of a fixed network infrastructure. LIME [15] defines an interesting and peculiar tuple-based architecture for handling in a uniform way both physical and actual agent mobility, also in the context of MANETs. It integrates useful forms of reactivity of the coordination media. However, LIME does not reach the full programmability required for context-dependent coordination and does not promote in any case an organizational perspective.

Again, the type of interaction model that one may wish to adopt for the virtual interaction media may strongly influence the complexity and the efficiency of the implementation. A peer-to-peer interaction model it is likely to limit the state size of the virtual interaction medium, and this it is like to reduce the problem of maintaining the consistency of the state of the medium and of properly enacting the coordination laws. However, as MANET typically defines a very dynamic scenario, in which agent can connect and disconnect very frequently from a group, relying on direct peer-to-peer agent interactions may be problematic. Conversely, a data-oriented model such as a tuple-based one, by fully uncoupling interacting entities, alleviates the problem related to the dynamism of agents' connection and disconnection. However, it introduces the problem of maintaining the consistency of a possibly very large state size of the coordination medium. In between, the adoption of an event-based model, exploiting a global data status of limited size, may achieve a good trade-off between the two issues of uncoupling and state consistency.

An additional issue that arises in MANET is that the lack of a fixed infrastructure makes the distinction between "local organizational laws" and "applications-specific organizational laws" less sharp. However, the detailed analysis of this issue would require investigating the possible application scenarios of MANET, and it is outside the scope of this paper.

4 Towards a General Model

The introduced organizational framework – together with the availability of a programmable coordination infrastructure – can provide advantages and reduce complexity of application design and development. However, they also introduce the issues – in the design phase – of identifying organizational rules and their composition and – in the development and maintenance phases – of correctly programming coordination laws into coordination media. More drastically, context-dependent coordination promotes a shift of focus from "engineering components" to "engineering organizational contexts" (or "engineering infrastructures").

In another work [2], we have shown several simple examples of dynamic and harmless composition of coordination laws in our MARS tuple-based programmable coordination infrastructure. However, in general, programming the behavior of the interaction space via complex compositions of coordination may be very difficult to be handled. In addition, local organizational laws may have to be dynamically composed with application-specific organizational laws. As a consequence, for the approach to be effective and practically usable there is the need for models of programmable interaction contexts, enabling a suitable and engineered approach for the definition and verification of coordination laws.

Unlike other computational models [3], the model should abstract away from internal details about the behavior of agents, e.g., of the entities that are intended to be coordinated by the model. Instead, it only has to focus on the *observable behavior* of components [5]: no matter what agents internally do, what it is important is that, as far as interactions with the external world are concerned, coordination laws can be effectively modeled and their properties verified. Such a property of the model is of dramatic importance in open scenario, as it is the case of the Internet and of mobile ad-hoc network, where one can neither assume any knowledge about the "internals" of the software components, nor act on it.

The starting point for the definition of our model is to conceive an interaction context as having an event-based space as a kernel:

- At the kernel-level, agents interact via the interaction context by generating events and by subscribing to events, that the event space handles according to the specific behavior programmed in it (Figure 3).
- At the application level, however, agents do not necessarily perceive the interaction context as an event-based space, nor they have to interact via a publish-subscribe event model (Figure 4).

In general, an event is characterized by a set of parameters and we indicate it by using the following notation: *E(x,y,z)*.

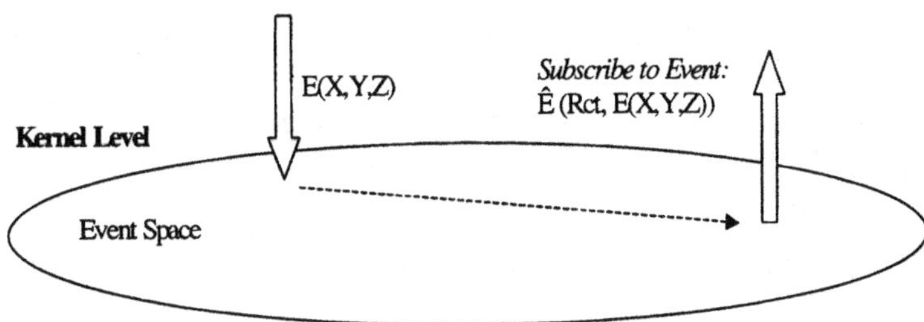

Fig. 3. Interaction Context: Kernel Level.

Instead the subscription mechanism consists in associating reactions to class of events. Reactions are associated to specific events on the basis of the parameters that characterize an event. We indicate an event subscription with the notation *Ê(Rct, E(x,y,z))*: it means that the reaction Rct (which can be an arbitrary piece of code) will be triggered when the event *E(x,y,z)* is fired. A subscription can have some non-defined values, in which case it associates the specified reaction to all the events that match it. For example, the subscription *Ê(Rct, E(null,y,null))* associates the reaction *Rct* to all the events described by three parameters, having *y* as second parameter. This approach is clearly inspired by our previous work on MARS [1]. Moreover, even events can have some non-defined values: e.g. *E(null,X)*, and subscriptions like *Ê(Rct, E(A,null))* are triggered by events like the previous one, because the *null* value in the event is matched with the *A* value of the subscription, while the *null* value of the subscription is matched with the *X* value of the event.

This publish-subscribe event model is very general and in the next section we will show how it can be exploited effectively.

4.1 Application-Level Events

Despite its kernel-level event-based architecture, the interaction context can be perceived and accessed by agents accordingly to any suitable model, e.g. a message-passing model or a tuple-based one. In order to interact and coordinate according to a given interaction model, agents typically invoke operations and do actions, that can be easily modeled in terms of events generations and/or subscriptions [22]. In other words, a given interaction model simply defines a sort of high-level operational interface to the event space (Figure 4). From an implementation point of view, we can imagine to have different coordination interfaces that map agents operations to events. For example a Linda-like coordination interface will translate typical Linda-like operations (i.e. *in, rd* and *out*) to events.

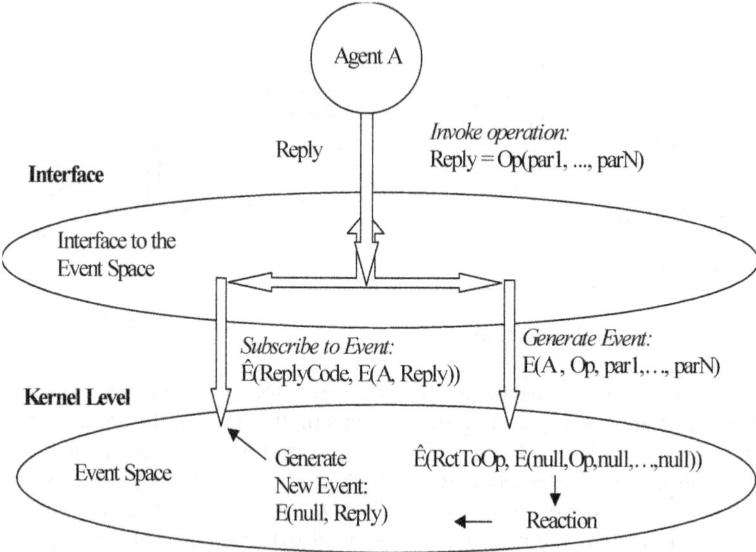

Fig. 4. Interaction Context: Application Level.

For each coordination interface, a set of subscriptions have to be preinstalled into the event kernel, in order to let the interaction model used (e.g. Linda) be realized in terms of the event model. In general these subscriptions have the form:

Ê (RctToOp, E(null, Op, null, ..., null))

So, in general (Figure 4), if agent *A* performs the operation *Op* with parameters $par_1,\ldots,par_n$:

$Op(par_1,\ldots.par_n)$

This generates an event on the interaction space, characterized by a set of parameters derived from the characteristics of the generating operation as follows:

$E(A, Op, par_1,\ldots,par_n)$

This matches the previous subscription and triggers the reaction RctToOp.

In addition, the operation may (but does not necessarily) imply the subscription of agent A for being notified about a reply event (Figure 4):

Ê (ReplyCode, E(A, Reply))

Let us clarify these concepts with an example. Consider an agent A, accessing the coordination media through a Linda-like interface, who performs the operation *out(tupleX)*. We already know that the tuple-based coordination interface will translate its request generating an event *E(A,out,tupleX)*. However there is also need that the event space has been previously programmed in order to let it store someway the tuple *tupleX*. When the Linda-like interface is installed upon the event-space, the following subscription is placed in the event-based space for dealing with the *out* operation:

Ê (code1, E(null,out,null))

The reaction *code1* is in charge to store the tuple identified by the third parameter of the event in the event space and to notify to possibly waiting agents the arrival of the new tuple.

Instead, if an agent B performs the Linda operation *rd(tupleX)*, the coordination interface has to install the subscription:

Ê (code3, E(B,tupleX))

It implies that when the reply event *E(B,tupleX)* is produced, then the reaction *code3* is executed, which returns the tuple *tupleX* to agent B. Then the coordination interface translates B's original request by the event *E(B,rd,tupleX)*.

Moreover, when the Linda-like interface is installed upon the event-space, the following reaction is placed in the event-based space:

Ê (code2, E(null,rd,null))

The reaction*code2* is in charge to search the tuple identified by the third parameter of the event, and if it is found, it generates the event *E(null, tuple-found)* e.g. *E(null, tupleX)*. The first parameter is set to *null* to specify that this event is catchable from all the agents made *rd(tupleX)*.

In this way when the event *E(null, tupleX)* is fired, it is matched with *Ê(code3, E(B,tupleX))* and *tupleX* is returned to agent B.

The event-space coordination space can serve not only as a coordination kernel, but also to deal with other agent aspects. In general, particular interfaces can translate to events agents' lifecycle actions, i.e., with regard to mobility, those related to the arrival and the departure of agents to and from interaction contexts. Again, these events can be characterized by different parameters (e.g., specifying the identity and the nature of the agent, its budget, etc.) However, it is worth emphasizing that modeling the arrival/departure of an agent in an interaction context in terms of events once again enforces a uniform and general way of handling mobility. In fact, from the point of view of the interaction context and of the events it perceives, there is no difference between the virtual, actual, or physical arrival of an agent. If necessary, specific event parameters can be used to associate different event handlers for different types of agents.

4.2 The Event-Kernel

The event space is the place in which all the events generated at the application level are caught and processed and where event subscriptions' pattern matching is executed. Programmed behaviors are represented by processes waiting to be triggered by the occurrence of specific events (or classes of events) to which they are subscribed. Programmed behaviors, once triggered, can generate new events, and subscribe to events.

In addition, the event space has a persistent state, and programmed behaviors have the capability of accessing the state of the event space, modifying it, and discriminate among their actions depending on the state. Events' subscriptions and reactions are stored in the event space persistent state. For example, the event space persistent state can be exploited by a tuple-based coordination interface as a mean to store the tuples written in the tuple space. If we take in consideration the example made in the previous section, the tuple-based coordination interface is in charge to program the behavior of the event kernel so that it will respond to the event *E(A,out,tupleX)* by storing the tuple *tupleX* in the persistent state. From the programmed behavior point of view, subscription to events occurs via an associative mechanism by specifying a "template" event. A matching mechanism checks the possibility of match between the template event and any event occurred. If there is match, the execution of the process is triggered to handle the event. In particular a pattern-matching mechanism is activated for any access to the event space, to detect the presence of reactions to be executed and trigger their execution. This is a very general mechanism to event subscription, also adopted in other event-based systems and models, e.g., JEDI [4].

Although our model can be effectively used to model interaction contexts build around different specific interaction models, what distinguishes it from other approaches to the modeling of interactions is that we do not fix the behavior of the event space in reacting to events. In non-programmable interaction spaces the behaviors in reaction to a given event is fixed once and for all by the model:

E → fixed_bhvr

In our approach, the behavior in response to a given event can be programmed:

E → programmed_bhvr

Moreover, programming of each interaction context is performed to embed in it both site-specific coordination laws and application-specific ones, so as to result in a dynamic composition of different behaviors, independently programmed by different entities and in different times:

E → prog_bhvr|prog_bhvr2|prog_bhvr3

This programmable model promotes a great dynamism in the event-reaction flow. In fact, if several templates satisfy the matching mechanism, all the corresponding reactions are executed according to their installation order. Moreover if the reaction to one event generates other events, these are matched for other possible reactions, and possibly other reactions can be fired (Figure 5). Because agents or site-administrators can add new reactions on the fly at any time, the reaction flow can change in a complete dynamic way. In fact, a single new reaction, which generates events, can in principle lead to a chain reaction that alter the global coordination media behavior.

At the time of writing, we are completing the definition of the formal and operational semantics of the model, and the evaluation of its expressiveness. The idea is that the operational semantics, once completed, can be effectively used whenever a given computation inside the event space evolves so as to implement the required coordination laws, given the generation or subscription to specific events at the application level. For instance, one can use it to determine whether the insertion of an application-specific coordination law into one interaction context can undermine the global behavior of the space, and viceversa. Moreover, because of the great dynamism of the model proposed a careful analysis is needed to prevent endless recursions in events' reactions.

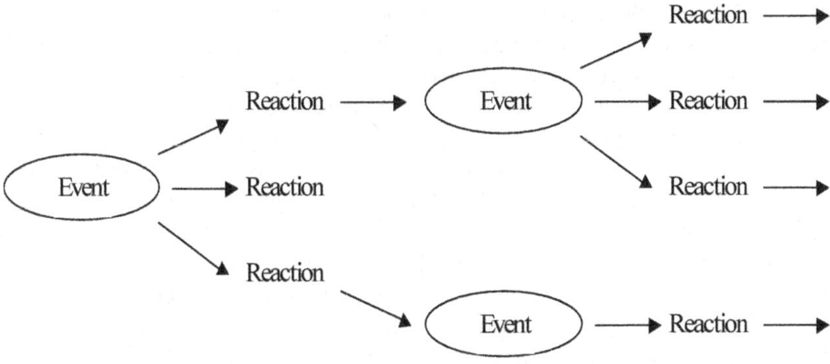

Fig. 5. Event-Reaction Flow.

5 Conclusions, Future Work, and Open Research Directions

In this paper we have argued that mobile application components can be effectively and uniformly modeled in terms of organizations of mobile and autonomous agents, and have introduced a conceptual framework for the modeling of such applications. When such a framework is supported, at the middleware level, by a proper programmable coordination infrastructure, it can make applications more modular and easier to develop and maintain. On this base, the paper has analyzed the main issues arising in the implementation of such a programmable infrastructure, also in the context of mobile ad-hoc networks. Moreover, we have outlined the need of a suitable formal model for the analysis of coordinated applications in mobile setting, and have proposed a general-purpose modeling of programmable infrastructure based on an event-based coordination kernel. The driving idea underlying both the analysis of the implementation issues and the proposal of the event-based model is that the effective design and development of mobile applications requires a careful engineering of the infrastructure and of its behavior, other than an engineering of the application components/agents.

Our current research work has two primary objectives.

On the one hand, we aim at completing the definition of the event-based model and at using it to analyze the characteristics of different coordination infrastructures, there included MARS, TuCSoN, and LIME. We also intend use the model to analyze and verify the properties of several proof-of-concepts infrastructures, with a specific attention to the MANET scenario. In addition, we will use the model to analyze different possible use-case coordination laws, enacting specific widely used coordination patterns or security policies.

On the other hand, building from both the results of the above analysis and our experience on the implementation of MARS, we plan to develop a coordination infrastructure built upon a programmable kernel implemented accordingly to the presented event-based model. The goal is to produce a system capable of coordinating mobile agents both with the support of a fixed network, wired or wireless, and also without any fixed infrastructure, in the context of MANETs. In this work, our main concerns will be scalability and interoperability. With regard to scalability, as mobility concepts gain success and more and more mobile agents – whether software and hardware – will populate the Internet and will forms MANETs, the capability of the infrastructure to scale well with the number of coordinated entities and its capability of distribute the burden between them will play a central role for acceptance and usability. With regard to interoperability, a key role will also be played by the capability of the infrastructure to take into consideration any existing standard as well as emerging ones. In this context XML is becoming the de-facto standard for data representation and exchange across the Internet. Therefore we wish to build a system completely XML compliant, not only for data exchange, but also for any coordination related issue. Moreover, we will look with particular attention to those new standards that aim to add semantic description to data, like DAML [25]. Only by exploiting these semantic capabilities, we could effectively manage interoperability between heterogeneous components.

During the above described research activity, we intend to assess and validate our results via the prototype development of real-world applications. We are particularly interested in application scenarios exploiting physical mobility, and we plan to develop applications targeted for wireless devices that incorporate Bluetooth or IEEE802.11 technologies.

Open research directions will take the moves for the following considerations.

We expect that in the next few years advances in wireless and embedded technologies will change completely the information technology application scenarios. On the one hand, new, small, powerful and intelligent components will be distributed and embedded everywhere in the environment and in our everyday objects (embedded systems). On the other hand, thanks to the advances in wireless communication, it will become less expensive and easy to connect all these systems with each other and to the Internet, and to integrate inherently mobile entities like appliances, robots, people, etc. This trend is already begun, and although the number of embedded and mobile devices is growing very fast, their number is still orders of magnitude below the next 20 years' expectations [17, 18]. We expect that the number of these components will somewhen reach an upper threshold over which radically new problems will arise, and conventional approaches and solutions will become inapplicable. In particular, we think it will not be possible to program and control these kinds of embedded applications in a conventional manner [24]. Our perception is that this kind of applications will require a deep exploitation of the notions of emergent behaviors and auto-adaptiveness. In particular we feel that the lessons of swarm-based computing [14], where engineering of complex – and useful – behaviors is achieved via environment-mediated interactions of a multitude of very simple agents, can be of much use in this direction. Thus, we plan to conduct simulations of large mobile applications, including thousands of mobile agents, with the goal of analyzing the global behaviors of these agents' ensembles, and of verifying how and to which extent a programmable coordination infrastructure can be use to control the global behavior of these large applications.

Acknowledgements

Work supported by the Italian MURST in the project "MUSIQUE – Infrastructure for QoS in Web Multimedia Services with Heterogeneous Access" and by the University of Modena and Reggio Emilia with a found for young researchers.

References

[1] G. Cabri, L. Leonardi, F. Zambonelli, "MARS: a Programmable Coordination Architecture for Mobile Agents", IEEE Internet Computing, 4(4), 2000.
[2] G. Cabri, L. Leonardi, F. Zambonelli, "Engineering Mobile Agent Applications via Context-Dependent Coordination", 23rd International Conference on Software Engineering, May 2001.
[3] L. Cardelli, A. D. Gordon, "Mobile Ambients", Theoretical Computer Science, 240(1), July 2000.
[4] G. Cugola, A. Fuggetta, E. De Nitto, "The JEDI Event-based Infrastructure", IEEE Transactions on Software Engineering, 2001, to appear.
[5] E. Denti, A. Natali, A. Omicini, "On the Expressive Power of a Language for Programmable Coordination Media", Proceedings of the 10th ACM Symposium on Applied Computing, ACM, 1998.
[6] T. Finin at al., "KQML as an Agent Communication Language", 3rd International Conference on Information Knowledge and Management", November 1994.
[7] A. Fuggetta, G. Picco, G. Vigna, "Understanding Code Mobility", IEEE Transactions on Software Engineering, 24(5), May 1998.
[8] D. Gelernter, N.Carriero "Coordination Languages and Their Significance", Communication of the ACM, Vol. 35, No. 2, pp. 96-107, February 1992.
[9] "T Spaces: the Next Wave", IBM System Journal, 37(3):454-474, 1998.
[10] N. R. Jennings, "On Agent-Based Software Engineering", Artificial Intelligence, 117(2), 2000.
[11] N.H. Minsky, V. Ungureanu, "Law-Governed Interaction: A Coordination & Control Mechanism for Heterogeneous Distributed Systems", ACM Transactions on Software Engineering and Methodology, 9(3), 2000.
[12] P. Noriega, C. Sierra, J. A. Rodriguez, "The Fishmarket Project. Reflections on Agent-mediated institutions for trustworthy E-Commerce", Workshop on Agent Mediated Electronic Commerce, 1998.
[13] A. Omicini, F. Zambonelli, "Coordination for Internet Application Development", Journal of Autonomous Agents and Multiagent Systems 2(3), Sept. 1999.
[14] V. Parunak, et al., "Distinguishing Environmental and Agent Dynamics", 1st International Workshop on engineering Societies in the Agents', Springer Verlag, LNCS 1972, 2000.
[15] G.P. Picco, A.M. Murphy, G. -C. Roman, "LIME: Linda Meets Mobility", 21st International Conference on Software Engineering, May 1999.
[16] G. P. Picco, A. M. Murphy, G. -C. Roman, "Software Engineering for Mobility: A Roadmap", in The Future of Software Engineering, A. Finkelstein (Ed.), ACM Press, pp. 241-258, 2000.
[17] D. Tennenhouse, "Proactive Computing", Communications of the ACM, 43(5): 43-50, May 2000.

[18] R. Tolskdorf, "Models of Coordination", 1st International Workshop on engineering Societies in the Agents', Springer Verlag, LNCS 1972, 2000.
[19] J. White, "Mobile Agents", in Software Agents, J. Bradshaw (Ed.), AAAI Press, pp. 437-472, 1997.
[20] F. Zambonelli, N. R. Jennings, A. Omicini, M. J. Wooldridge, "Agent-Oriented Software Engineering for Internet Applications", in Coordination of Internet Agents, A. Omicini, F. Zambonelli, M. Klusch, R. Tolksdorf (Eds.), Springer-Verlag, 2001.
[21] F. Zambonelli, N. R. Jennings, M. J. Wooldridge, "Organizational Abstractions for the Analysis and Design of Multi-agent Systems", in Agent-Oriented Software Engineering, Springer Verlag, LNCS 1947, 2000.
[22] G. Zavattaro, N. Busi, "Publish/Subscribe vs. Shared Dataspace Coordination Infrastructures", 10th IEEE Workshops on Enabling Technologies: Infrastructures for Collaborative Enterprises, IEEE CS Press, Boston (MA), June 2001.
[23] Paul Davidson, "Categories of Artificial Societies", Springer 2001, LNAI 2203, pp. 1-9.
[24] Martin Fredriksson, Rune Gustavsson, "A Methodological Perspective on Engineering Agent Societies", Springer Verlag, LNAI 2203, 2001, pp. 10-24.
[25] Filip Perich, Lalana Kagal, Harry Chen, Sovrin Tolia, Youyong Zou, Tim Finin, Anupam Joshi, Yun Peng, R. Scott Cost, Charles Nicholas, "ITTALKS: An Application of Agents in the Semantic Web", Springer Verlag, LNAI 2203, 2001, pp. 175-193.

Preferring and Updating in Abductive Multi-agent Systems

Pierangelo Dell'Acqua[1] and Luís Moniz Pereira[2]

[1] Department of Science and Technology, Campus Norrköping
Linköping University, Norrköping, Sweden
pier@itn.liu.se
[2] Centro de Inteligência Artificial - CENTRIA
Departamento de Informática, Faculdade de Ciências e Tecnologia
Universidade Nova de Lisboa, 2829-516 Caparica, Portugal
lmp@di.fct.unl.pt

Abstract. We present a logical framework and the declarative semantics of a multi-agent system in which each agent can communicate with and update other agents, can react to the environment, is able to prefer, whether beliefs or reactions, when several alternatives are possible, and is able to abduce hypotheses to explain observations. The knowledge state of an agent is represented by an updatable prioritized abductive logic program, in which priorities among rules can be expressed to allow the agent to prefer. We sketch two examples to illustrate how our approach functions, including how to prefer abducibles to tackle the problem of multiple hypotheses and how to perform the interplay between planning and acting.
We argue that the theory of the type of agents considered is a rich evolvable basis, and suitable for engineering configurable, dynamic, self-organizing and self-evolving agent societies.

1 Introduction

We present a logical formalization of a framework for multi-agent systems and we define its semantics. In this framework, we embed a flexible and powerful kind of agent. In fact, these agents are rational, reactive, abductive, able to prefer and they can update the knowledge base of other agents (including their own).

The knowledge state of each agent is represented by an an abductive logic program in which it is possible to express rules, integrity constraints, active rules and priorities among rules. This allows the agents to reason, to react to the environment, to prefer among several alternatives, to update both beliefs and reactions, and to abduce hypotheses to explain observations. We present a declarative semantics of this kind of agent.

These agents are then embedded into a multi-agent system in such a way that the only form of interaction among them is based on the notions of project and update. A project of the form $\alpha{:}C$ of an agent β denotes the intention of β of proposing to update the theory of an agent α with C. Correspondingly, an update of the form $\beta{\div}C$ in the theory of α denotes the intention of β to update the current theory of α with C. It is then up to α whether or not to accept that update. For example, if α trusts β and therefore α

A. Omicini, P. Petta, and R. Tolksdorf (Eds.): ESAW 2001, LNAI 2203, pp. 57–73, 2001.

is willing to accept it, then α has to update its theory with C. The new information may contradict what α believes and, if so, the new believed information will override what is currently believed by α. β can also propose an update to itself by issuing an internal project $\beta{:}C$.

The semantics of the multi-agent system provides a logical description of the agent interactions. The definition of the semantics depends also on the goal that each agent has to prove at a certain moment in time. In fact, by proving a goal the agent may abduce hypotheses that explain that goal, and in turn these hypotheses may trigger reactive rules, and so on. Hypotheses abduced in proving a goal G are not permanent knowledge, rather they only hold during the proof of G. To make them permanent knowledge, an agent can issue an internal project and update its own knowledge base with those hypotheses.

The engineering of our agent society is based exclusively on the notions of projects and updates to model the agent interactions since we aim at building autonomous and distributed agent systems. This form of interaction is powerful and flexible, and a number of communication protocols can be built based on it. However, the main contribution of the paper is to assemble all the ingredients of the agent architecture needed to engineer configurable, dynamic agent societies where the agents can self-organize themselves with respect to their goals, and self-evolve. In this way, the overall "emerging" structure will be flexible and dynamic: each agent will have its own representation of its organization which, furthermore, is updatable by preferring.

The remainder of the paper is structured as follows. Section 2 presents the logical framework. Section 3 introduces our conception of abductive agent and multi-agent system. The declarative semantics of abductive agents and multi-agent systems is presented in Sections 4 and 5, respectively. Two examples to illustrate how the approach functions, including how to prefer abducibles to tackle the problem of multiple hypotheses and how to perform the interplay between planning and acting, are shown in Sections 6 and 7. Finally, Section 8, Conclusion and Future Work, argues that the theory of the type of agents considered is a rich evolvable basis, and suitable for engineering configurable, dynamic, self-organizing and self-evolving agent societies, and discusses future work.

2 Logic Programming Framework

Typically, an agent can hold positive and negative information, and it can update its own knowledge with respect to the new incoming information. Thus the language of an agent should be expressive enough to represent both positive and negative information. In order to represent negative information in logic programs, we need a language that allows default negation $not\ A$ not only in premises of clauses but also in their heads [1]. We call such programs generalized logic programs. It is convenient to syntactically represent generalized logic programs as propositional Horn theories. In particular, we represent default negation $not\ A$ as a standard propositional variable.

[1] For further motivation and intuitive reading of logic programs with default negations in the heads see [2].

Propositional variables whose names do not begin with "not" and do not contain the symbols ":" and "÷" are called *objective atoms*. Propositional variables of the form $not\ A$ are called *default atoms*. Objective atoms and default atoms are generically called *atoms*.

Propositional variables of the form $\alpha{:}C$ (where C is defined below) are called *projects*. $\alpha{:}C$ denotes the intention (of some agent β) of proposing the updating the theory of agent α with C. Projects can be negated. A *negated project* of the form $not\ \alpha{:}C$ denotes the intention of the agent of not proposing the updating of the theory of agent α with C.

Propositional variables of the form $\beta \div C$ are called *updates*. $\beta \div C$ denotes an update that has been proposed by β of the current theory (of some agent α) with C. Updates can be negated. A *negated update* of the form $not\ \beta{\div}C$ in the theory of an agent α indicates that agent β does not have the intention to update the theory of agent α with C.

Definition 1. Let $\mathcal{K}$ be a set of propositional variables consisting of objective atoms and projects such that *false* $\notin \mathcal{K}$. The propositional language $\mathcal{L}_\mathcal{K}$ generated by $\mathcal{K}$ is the language which consists of the following set of propositional variables:

$$\begin{aligned}\mathcal{L}_\mathcal{K} = {} & \mathcal{K} \cup \{\mathit{false}\} \cup \{not\ A \mid \text{for every objective atom } A \in \mathcal{K}\} \\ & \cup \{not\ \alpha{:}C, \alpha{\div}C, not\ \alpha{\div}C \mid \text{for every project } \alpha{:}C \in \mathcal{K}\}.\end{aligned}$$

Definition 2 (Generalized Rule). A generalized rule in the language $\mathcal{L}_\mathcal{K}$ is a rule of the form: $L_0 \leftarrow L_1 \wedge \ldots \wedge L_n$ $(n \geq 0)$, where L_0 (with $L_0 \neq \mathit{false}$) is an atom and every L_i $(1 \leq i \leq n)$ is an atom, an update or a negated update from $\mathcal{L}_\mathcal{K}$.

Note that, according to the above definition, only objective atoms and default atoms can occur in the head of generalized rules.

We use the following convention. Given a generalized rule r of the form $L_0 \leftarrow L_1 \wedge \ldots \wedge L_n$, we use $head(r)$ to indicate L_0, $body(r)$ to indicate the conjunction $L_1 \wedge \ldots \wedge L_n$, $body_{pos}(r)$ to indicate the conjunction of all objective atoms and updates in $body(r)$, and $body_{neg}(r)$ to indicate the conjunction of all default atoms and negated updates in $body(r)$. Whenever L is of the form $not\ A$, $not\ L$ stands for the atom A.

Definition 3 (Integrity Constraint). An integrity constraint in the language $\mathcal{L}_\mathcal{K}$ is a rule of the form: $\mathit{false} \leftarrow L_1 \wedge \ldots \wedge L_n \wedge Z_1 \wedge \ldots \wedge Z_m$ $(n \geq 0, m \geq 0)$, where every L_i $(1 \leq i \leq n)$ is an atom, an update or a negated update, and every Z_j $(1 \leq j \leq m)$ is a project or a negated project from $\mathcal{L}_\mathcal{K}$.

Integrity constraints are rules that enforce some condition over the state, and therefore always take the form of denials, without loss of generality, in a 2-valued semantics. Note that generalized rules are distinct from integrity constraints and should not be reduced to them. In fact, in generalized rules it is of crucial importance, when updating which atom occurs in the head.

Definition 4 (Query). A query Q in the language $\mathcal{L}_\mathcal{K}$ takes the form: $?{-}\ L_1 \wedge \ldots \wedge L_n$ $(n \geq 1)$, where every L_i $(1 \leq i \leq n)$ is an atom, an update or a negated update from $\mathcal{L}_\mathcal{K}$.

The following definition introduces rules that are evaluated bottom-up. To emphasize this aspect, we employ a different notation for them.

Definition 5 (Active Rule). An active rule in the language $\mathcal{L}_\mathcal{K}$ is a rule of the form: $L_1 \wedge \ldots \wedge L_n \Rightarrow Z$ $(n \geq 0)$, where every L_i $(1 \leq i \leq n)$ is an atom, an update or a negated update, and Z is a project or a negated project from $\mathcal{L}_\mathcal{K}$.

We use the following convention: given an active rule r of the form $L_1 \wedge \ldots \wedge L_n \Rightarrow Z$, we use $head(r)$ to indicate Z, and $body(r)$ to indicate $L_1 \wedge \ldots \wedge L_n$.

Active rules are rules that can modify the current state, to produce a new state, when triggered. If the body $L_1 \wedge \ldots \wedge L_n$ of the active rule is satisfied, then the project (fluent) Z can be selected and executed. The head of an active rule must be a project that is either internal or external. An *internal project* operates on the state of the agent itself (self-update), e.g., if an agent gets an observation, then it updates its knowledge, or if some conditions are met, then it executes some goal. *External projects* instead are performed on the environment, e.g., when an agent wants to update the theory of another agent. A negated project that occurs in the head of an active rule denotes the intention (of some agent) not to perform that project at the current state.

We assume that for every project $\alpha{:}C$ in $\mathcal{K}$, C is either a generalized rule, an integrity constraint, an active rule or a query. Thus, a project can only take one of the following forms:

$$\alpha{:}(L_0 \leftarrow L_1 \wedge \ldots \wedge L_n) \qquad \alpha{:}(L_1 \wedge \ldots \wedge L_n \Rightarrow Z)$$
$$\alpha{:}(\mathit{false} \leftarrow L_1 \wedge \ldots \wedge L_n \wedge Z_1 \wedge \ldots \wedge Z_m) \qquad \alpha{:}(?{-}L_1 \wedge \ldots \wedge L_n)$$

Note that projects and negated projects can only occur in the head of active rules and in the body of integrity constraints.

Example 1. The integrity constraint $\mathit{false} \leftarrow A \wedge \beta{:}B$ in the theory of an agent α enforces the condition that α cannot perform a project $\beta{:}B$ when A holds. The active rule $A \wedge \mathit{not}\,\beta{\div}B \Rightarrow \beta{:}C$ in the theory of an agent α instructs it to perform project $\beta{:}C$ if A holds and agent β has not wanted to update the theory of α with B.

Let $<$ be a binary predicate symbol whose set of constants includes all the generalized rules in the language $\mathcal{L}_\mathcal{K}$.

Definition 6 (Priority Rule). A priority rule in the language $\mathcal{L}_\mathcal{K}$ is a generalized rule defining the predicate symbol $<$.

It is assumed that the set of constants of $<$ does not include $<$ itself. $r_1 < r_2$ means that rule r_1 is preferred to rule r_2.

Definition 7 (Prioritized Logic Program). A prioritized logic program P in the language $\mathcal{L}_\mathcal{K}$ is a set of generalized rules (possibly, priority rules) and integrity constraints in the language $\mathcal{L}_\mathcal{K}$.

Definition 8 (Prioritized Abductive Logic Program). A prioritized abductive logic program is a pair $(P, \mathcal{A})$, where P is a prioritized logic program and $\mathcal{A}$ is a set of atoms in the language $\mathcal{L}_\mathcal{K}$. The atoms in $\mathcal{A}$ are referred to as the abducibles.

The definition of prioritized abductive logic program generalizes the one given in [9] to allow for priority rules. The intuition is that the program P formalizes the theory of an agent, and the priority rules express preferences among rules in P, in a sense to be made concrete later. Abducibles can be thought of as hypotheses that can be used to extend the given prioritized logic program in order to provide an "explanation" for given queries. Explanations are required to "satisfy" the integrity constraints in P. Abducibles may also be defined in P by generalized rules as the result of a self-update which adopts an abducible as a fact.

3 Abductive Agents and Multi-agent Systems

This section presents the conception of abductive agent and of multi-agent system. The initial knowledge of an agent is modeled by the notion of initial theory.

Definition 9 (Initial Theory). The initial theory $\mathcal{T}$ of an agent α is a tuple $(P, \mathcal{A}, R)$, where $(P, \mathcal{A})$ is a prioritized abductive logic program and R is a set of active rules.

$(P, \mathcal{A})$ formalizes the initial knowledge state of the agent, and R characterizes its reactive behaviour. The knowledge of an agent can dynamically evolve when the agent receives new knowledge, albeit by self-updating rules, or when it abduces new hypotheses to explain observations. The new knowledge is represented in the form of an updating program, and the new hypotheses in the form of a (finite) set $\triangle \subseteq \mathcal{A}$ of abducibles, possibly negated.

Definition 10 (Updating Program). Let $\mathcal{M}$ be a multi-agent system (defined below). An updating program U is a finite set of updates such that if an update $v{\div}C \in U$ then v is an agent of $\mathcal{M}$.

An updating program contains the updates that will be performed on the current knowledge state of the agent. To characterize the evolution of the knowledge of an agent we need to introduce the notion of sequence of updating programs. Let $S = \{0, 1, \ldots, m\}$ be a set of natural numbers. We call the elements $s \in S$ *states*. A *sequence of updating programs* $\mathcal{U} = \{U^s \mid s \in S \text{ and } s > 0\}$ is a set of updating programs U^s superscripted by the states $s \in S$.

Definition 11 (Agent α at State s). Let $s \in S$ be a state. An agent α at state s, written as Ψ^s_α, is a pair $(\mathcal{T}, \mathcal{U})$, where $\mathcal{T}$ is the initial theory of α and $\mathcal{U} = \{U^1, \ldots, U^s\}$ is a sequence of updating programs. If $s = 0$, then $\mathcal{U} = \{\}$.

An agent α at state 0 is defined by its initial theory and by an empty sequence of updating programs, that is $\Psi^0_\alpha = (\mathcal{T}, \{\})$. At state 1, α is defined by $(\mathcal{T}, \{U^1\})$, where U^1 is the updating program containing all the updates that α has received at state 1, either from other agents or as self-updates. In general, an agent α at state s is defined by $\Psi^s_\alpha = (\mathcal{T}, \{U^1, \ldots, U^s\})$, where each U^i is the updating program containing the updates that α has received at state i.

Definition 12 (Multi-agent System at State s). A multi-agent system $\mathcal{M} = \{\Psi^s_{\alpha_1}, \ldots, \Psi^s_{\alpha_n}\}$ at state s is a set of agents $\alpha_1, \ldots, \alpha_n$ each of which at state s.

Note that the definition of multi-agent system characterizes a static society of agents in the sense that it is not possible to add/remove agents from the system, and all the agents are at one common state. Distinct agents in $\mathcal{M}$ may have different sets $\mathcal{A}$ of abducibles.

Within logic programs we refer to agents by using the corresponding subscript. For instance, if we want to express the update of the theory of an agent Ψ_α with C, we write the project $\alpha{:}C$.

To begin with, the system starts at state 0 where each agent α is defined by $\Psi^0_\alpha = (\mathcal{T}_\alpha, \{\})$. Suppose that at this state α abduces $\triangle = \{a, \mathit{not}\ b\}$ to provide an explanation for an observation made. Suppose also that at state 0 agent β wants to propose an update of the knowledge state of agent α with C by triggering the project $\alpha : C$. Then, at the next state α will receive the update $\beta \div C$ indicating that an update has been proposed by β. Thus, at state 1, α will be defined by $(\mathcal{T}_\alpha, \{U^1\})$, where $U^1 = \{\beta \div C\}$ if no other updates occur for α.

The hypotheses $\triangle$ abduced by α at state 0 are by default discarded at state 1. Thus, α normally does not have memory of what it has assumed. If we want instead to model an agent α that is able to enrich its experience by assuming hypotheses during the process of proving queries and explaining observations, and it is able to adopt previously assumed hypotheses, we can equip the theory of α with active rules of the form: $a \Rightarrow \alpha{:}a$, for whatever abducible $a \in \mathcal{A}$ desired. In this way, when α abduces a the project $\alpha{:}a$ will be triggered, and at the next state its theory will be updated with a itself.

4 Semantics of Abductive Agents

This section introduces the declarative semantics of abductive agents and of multi-agent systems. We need the following definitions. Let U be an updating program.

$$\Pi(U) = \{r \mid r \text{ is a generalized rule and } \alpha{\div}r \in U\}$$
$$\Omega(U) = \{a \mid a \text{ is an active rule and } \alpha{\div}a \in U\}$$

Definition 13 (Default Assumptions). Let $(P, \mathcal{A})$ be a prioritized abductive logic program and M a model of P. Let $\triangle \subseteq \mathcal{A}$ be a set of abducibles. The set of default assumptions[2] is:

$$\mathit{Default}(P, \triangle, M) = \{\mathit{not}\ A \mid A \text{ is an object atom}, A \notin \triangle, \mathit{not}\ A \notin \triangle, \text{ and } \nexists r \in P \text{ such that } \mathit{head}(r) = A \text{ and } M \models \mathit{body}(r)\}.$$

The knowledge of an agent α is characterized at the start by its initial theory $\mathcal{T}$. Its knowledge can dynamically evolve when α receives new knowledge, via a sequence of updating programs $\mathcal{U} = \{U^1, \ldots, U^s\}$, or when α assumes new hypotheses to explain new observations, via a set of abducibles $\triangle$. Intuitively, the evolution of knowledge may be viewed as the result of, starting with $\mathcal{T}$, updating it with U^1, updating next with U^2, and so on. Similarly to the rationale of Dynamic Logic Programming (DLP) [2], the rules proposed via updates can be added to the knowledge state of α provided

[2] For simplicity, we assume positive abducibles false by default.

that α does not distrust the update, for instance if it goes against its presently assumed hypotheses and without bothering whether they conflict with previous knowledge of α itself. For example, if an agent α receives an update of the form $\beta \div C$ and α does not distrust it, then the rule C will be added to the knowledge state of α. Also similarly to DLP, the role of updating is to ensure that the rules contained in these newly added updates are in force, and that previous rules are still valid (by inertia) as far as possible, i.e., they are in force while they do not conflict with newly added rules, and they lose force if they so conflict and the new rules remain in force themselves. This rationale is at the basis of the notion of rejected rules.

Definition 14 (Rejected Generalized Rule). Let $(P, \mathcal{A})$ be a prioritized abductive logic program and M a model of P. Let $s \in S$ be a state of an agent, $\mathcal{U} = \{U^i \mid i \in S \text{ and } i > 0\}$ a sequence of updating programs and $\Delta \subseteq \mathcal{A}$ a set of abducibles. The set of rejected generalized rules at state s is:

$$
\begin{aligned}
&\mathit{RejectGr}(P, \mathcal{U}, \Delta, s, M) = \\
&\{r \in P \mid r \text{ is a generalized rule and } \exists\, \alpha \div r' \in U^i \text{ such that } 1 \leq i \leq s, \\
&\qquad \mathit{head}(r) = \mathit{not}\ \mathit{head}(r'), M \models \mathit{body}(r') \text{ and } M \models \mathit{not}\ \mathit{distrust}(\alpha \div r')\} \\
&\cup \{r \in P \mid r \text{ is a generalized rule and } \exists\, L \in \Delta \text{ such that } \mathit{head}(r) = \mathit{not}\ L\} \\
&\cup \{r \mid r \text{ is a generalized rule, } \exists\, \beta \div r \in U^i \text{ and } \exists\, \alpha \div r' \in U^j \text{ such that } i < j \leq s, \\
&\qquad \mathit{head}(r) = \mathit{not}\ \mathit{head}(r'), M \models \mathit{body}(r') \text{ and } M \models \mathit{not}\ \mathit{distrust}(\alpha \div r')\} \\
&\cup \{r \mid r \text{ is a generalized rule, } \exists\, \beta \div r \in U^i, 1 \leq i \leq s \text{ and } \exists\, L \in \Delta \text{ such that} \\
&\qquad \mathit{head}(r) = \mathit{not}\ L\}.
\end{aligned}
$$

According to the definition above, a generalized rule r either in P or proposed via an update in U^i is rejected at state s by a model M if there exists a newer generalized rule r' proposed via a subsequent update in U^j by any agent α, such that the head of r' is the complement of the head of r, the body of r' is true in M and the update is not distrusted. r can also be rejected if there exists an hypothesis $L \in \Delta$ that is the complement of the head of r. Mark that: (1) whenever an agent distrust an update proposed by another agent, the rule proposed is not in force; (2) at state 0 no rule is rejected, i.e., $\mathit{RejectGr}(P, \mathcal{U}, \Delta, 0, M) = \{\}$; (3) all non rejected rules persist by inertia; (4) *distrust/1* is a reserved predicate which can itself be updated; (5) it is not required of r' itself not to be rejected to allow it to reject r, simply because if r' is itself rejected then some subsequent rule has reinstated the conclusion of r and so the rejection of r is immaterial; this comment applies as well to the next definition. Generalized rules defining abducibles can also be rejected. Abducibles may have defining rules in the theory of an agent α either because communicated by other agents via updates, or because of a self-update, like in the case where there exists an active rule of the form $a \Rightarrow \alpha : a$ (with $a \in \mathcal{A}$). Thus, an agent may assume new hypotheses to motivate an observation, and later on discard those hypotheses because it turns out (by performing subsequent observations) to be not correct or because the world has evolved and those hypotheses became obsolete.

Active rules can also be rejected in a way similar to that of generalized rules.

Definition 15 (Rejected Active Rule). Let $(P, \mathcal{A}, R)$ be the initial theory of an agent and M a model of P. Let $s \in S$ be a state and $\mathcal{U} = \{U^i \mid i \in S \text{ and } i > 0\}$ a sequence of updating programs. The set of rejected active rules at state s is:

$$\begin{array}{l} RejectAr(R, \mathcal{U}, s, M) = \\ \{r \in R \mid \exists\, \alpha{\div}r' \in U^i \text{ such that } 1 \leq i \leq s, head(r) = not\ head(r'), \\ \qquad M \models body(r') \text{ and } M \models not\ distrust(\alpha{\div}r')\} \\ \cup\, \{r \mid \exists\, \beta{\div}r \in U^i, r \text{ is an active rule and } \exists\, \alpha{\div}r' \in U^j \text{ such that } i < j \leq s, \\ \qquad head(r) = not\ head(r'), M \models body(r') \text{ and } M \models not\ distrust(\alpha{\div}r')\} \end{array}$$

As the head of an active rule is a project and not an atom, active rules can only be rejected by active rules. Rejecting an active rule r makes r not triggerable even if its body is true in the model. Thus, by rejecting active rules, we make the agent less reactive.

While updates allow us to deal with a dynamically evolving world, where rules change in time, preferences allow us to choose among various possible models of the world and among possible incompatible reactions. In [4], two criteria are established to remove unpreferred generalized rules in a program: removing unsupported generalized rules, and removing less preferred generalized rules defeated by the head of some more preferred one. Unsupported generalized rules are rules whose head is true in the model and whose body is defeated by the model.

Definition 16 (Unsupported Generalized Rule). Let P be a prioritized logic program and M a model of P. The set of unsupported generalized rules is:

$$Unsup(P, M) = \{r \in P \mid M \models head(r), M \models body_{pos}(r) \text{ and } M \not\models body_{neg}(r)\}.$$

Definition 17 (Unpreferred Generalized Rule). Let P be a prioritized logic program and M a model of P. $Unpref(P, M)$ is a set of unpreferred generalized rules of P and M iff:

$$Unpref(P, M) = \text{least}(Unsup(P, M) \cup \mathcal{X})$$

where $\mathcal{X} = \{\ r \in P \mid \exists r' \in (P - Unpref(P, M))$ such that:
$$M \models r' < r,\ M \models body_{pos}(r') \text{ and } [\ not\ head(r') \in body_{neg}(r) \text{ or } (not\ head(r) \in body_{neg}(r'),\ M \models body(r))\ ]\ \}.$$

In other words, a generalized rule is unpreferred if it is unsupported or defeated by a more preferred generalized rule (which is not itself unpreferred), or if it attacks (i.e., attempts to defeat) a more preferred generalized rule.

Active rules are rules that enjoy the property that they can be triggered when their body is true in the model. Triggering an active rule means to execute the project occurring in its head. For example, given an active rule $L_1 \wedge L_2 \Rightarrow \alpha : C$ and a model M, if $M \models L_1 \wedge L_2$, then the project $\alpha : C$ can be executed. Executing the project $\alpha : C$ denotes the intention of the agent of updating the knowledge state of agent α with C. The set of executable projects is the set containing the projects of all the active rules that are triggered and preferred by a model M.

Definition 18 (Executable Project). Let R be a set of active rules and M a model. The set of executable projects is:

$$Project(R, M) = \{\alpha{:}C \mid \exists r \in R \text{ such that } head(r) = \alpha{:}C \text{ and } M \models body(r)\}.$$

The following definition introduces the notion of preferred abductive stable model of an agent α at a state s with set of hypotheses $\triangle$. Given the initial theory $\mathcal{T}$ of α, a sequence of updating programs $\mathcal{U}$ and the hypotheses $\triangle$ assumed at state s by α, a preferred abductive stable model of α at state s is a stable model of the program $\mathcal{X}$ that extends P to contain all the updates in $\mathcal{U}$, all the hypotheses in $\triangle$, and all those rules whose updates are not distrusted but that are neither rejected nor unpreferred. The preferred abductive stable model contains also the projects of all active rules (both in P or proposed via an update) that are triggered (i.e., all the executable projects).

Definition 19 (Preferred Abductive Stable Model of Agent α at State s with Hypotheses $\triangle$). Let $s \in S$ be a state. Let $\Psi_\alpha^s = (\mathcal{T}, \mathcal{U})$ be agent α at state s and M an interpretation such that $false \notin M$. Assume that $\mathcal{T} = (P, \mathcal{A}, R)$ and $\mathcal{U} = \{U^i \mid i \in S \text{ and } i > 0\}$. Let $\triangle \subseteq \mathcal{A}$ be a set of abducibles. M is a preferred abductive stable model of agent α at state s with hypotheses $\triangle$ iff:

– $\forall r_1, r_2$: if $(r_1 < r_2) \in M$, then $(r_2 < r_1) \notin M$

– $\forall r_1, r_2, r_3$: if $(r_1 < r_2) \in M$ and $(r_2 < r_3) \in M$, then $(r_1 < r_3) \in M$

– $M = \text{least}(\mathcal{X} \cup Default(\mathcal{Y}, \triangle, M) \cup Project(\mathcal{Z}, M))$, where:

$$\mathcal{Y} = P \cup \bigcup_{1 \le i \le s} U^i \cup \triangle \cup \{r \mid r \text{ is a generalized rule or an integrity constraint and } \exists\, \alpha \div r \in \bigcup_{1 \le i \le s} U^i \text{ such that } M \models not\ distrust(\alpha \div r)\}$$

$$\mathcal{X} = \mathcal{Y} - RejectGr(P, \mathcal{U}, \triangle, s, M) - Unpref(\mathcal{Y} - RejectGr(P, \mathcal{U}, \triangle, s, M), M)$$

$$\mathcal{Z} = R \cup \{r \mid r \text{ is an active rule and } \exists\, \alpha \div r \in \bigcup_{1 \le i \le s} U^i\} - RejectAr(R, \mathcal{U}, s, M).$$

In general, at a certain state an agent α may have several preferred abductive stable models. Note that if $\triangle = \{\}$ and $s = 0$ or each updating program in $\mathcal{U}$ is empty, then the last condition of the definition above reduces to:

$$\mathcal{Y} = P \qquad \mathcal{X} = \mathcal{Y} - Unpref(\mathcal{Y}, M)$$
$$M = \text{least}(\mathcal{X} \cup Default(\mathcal{Y}, \{\}, M) \cup Project(R, M)).$$

This condition reduces further, if (in addition) there are no preference rules in P, to:

$$\mathcal{Y} = P \qquad \mathcal{X} = \mathcal{Y} - Unsup(\mathcal{Y}, M)$$
$$M = \text{least}(\mathcal{X} \cup Default(\mathcal{Y}, \{\}, M) \cup Project(R, M))$$
$$= \text{least}(P \cup Default(\mathcal{Y}, \{\}, M) \cup Project(R, M)).$$

When there are neither projects nor updates nor hypotheses (i.e., $\triangle = \{\}$), the semantics reduces to the Preferential semantics of Brewka and Eiter [6]. If updates are then introduced, the semantics generalizes to the Updates and Preferences semantics of Alferes and Pereira [4], which extends DLP with preferences. Our semantics takes the latter and complements it with abducibles, with mutual and self-updates by means of active rules and projects, plus queries, within a society of agents.

Definition 20 (Abductive Explanation of Agent α at State s for Query Q). Let $s \in S$ be a state and $\Psi_\alpha^s = (\mathcal{T}, \mathcal{U})$ agent α at state s. Let Q be a query. An abductive explanation of agent α at state s for Q is any subset $\triangle$ of $\mathcal{A}$ such that there exists a preferred abductive stable model M of α at s with hypotheses $\triangle$ and $M \models Q$.

5 Semantics of Multi-agent Systems

We can now present the semantics $\mathcal{S}$ of a multi-agent system $\mathcal{M}$ embedding abductive agents. The role of $\mathcal{S}$ is to characterize the relationship among the agents of $\mathcal{M}$. This is achieved by formalizing the relation between projects and updates. $\mathcal{S}$ is defined as the set of preferred abductive stable models of each agent of $\mathcal{M}$. It uses the following function. Let $\{M^{s-1}_{\alpha_1}, \ldots, M^{s-1}_{\alpha_n}\}$ be a set of the set of models of the agents $\alpha_1, \ldots, \alpha_n$ at state $s-1$.

$$\gamma(\{M^{s-1}_{\alpha_1}, \ldots, M^{s-1}_{\alpha_n}\}) = \{U^s_{\alpha_1}, \ldots, U^s_{\alpha_n}\}$$

where an update $(\alpha_j \div C) \in U^s_{\alpha_i}$ iff the project $\alpha_i{:}C$ belongs to all the models of α at state $s-1$, that is: $(\alpha_i{:}C) \in \bigcap M^{s-1}_{\alpha_j}$, for $1 \leq i, j \leq n$. The intuition is that α at state s can have several preferred abductive stable models, each of which may trigger distinct active rules (and therefore each model will contain distinct executable projects). According to γ the projects that will be executed are the selectable projects that occur in every model of α at state s. Note that when $j = i$ we have a self-update: the agent chooses to update its own theory.

Definition 21 (Semantics of $\mathcal{M}$). Let $\mathcal{M} = \{\Psi^0_{\alpha_1}, \ldots, \Psi^0_{\alpha_n}\}$ be an abductive multi-agent system at state 0. Let $s \in S$ be a state. The semantics $\mathcal{S}^s$ of $\mathcal{M}$ at state s is defined inductively as follows.

Base Case ($s = 0$) That means that $\Psi^0_{\alpha_i} = (\mathcal{T}_{\alpha_i}, \{\})$, for every agent α_i (with $1 \leq i \leq n$). Suppose that $Q^0_{\alpha_i}$ is the query of α_i at state 0. Let $M^0_{\alpha_i}$ be the set of preferred abductive stable models of α_i at state 0 each of which with hypothesis any abductive explanation $\Delta^0_{\alpha_i}$ of α_i for $Q^0_{\alpha_i}$. Then, the semantics $\mathcal{S}^o$ of $\mathcal{M}$ is $\mathcal{S}^0 = \{M^0_{\alpha_1}, \ldots, M^0_{\alpha_n}\}$.

Inductive step ($s > 0$) Let $\{\Psi^{s-1}_{\alpha_1}, \ldots, \Psi^{s-1}_{\alpha_n}\}$ be the multi-agent system $\mathcal{M}$ at state $s-1$ and $\mathcal{S}^{s-1} = \{M^{s-1}_{\alpha_1}, \ldots, M^{s-1}_{\alpha_n}\}$ its semantics. Suppose that $\Psi^{s-1}_{\alpha_i} = (\mathcal{T}_{\alpha_i}, \{U^1_{\alpha_i}, \ldots, U^{s-1}_{\alpha_i}\})$, for $1 \leq i \leq n$. Assume that $\gamma(\mathcal{S}^{s-1}) = \{U^s_{\alpha_1}, \ldots, U^s_{\alpha_n}\}$.
Then, the multi-agent system $\mathcal{M}$ at state s is $\{\Psi^s_{\alpha_1}, \ldots, \Psi^s_{\alpha_n}\}$ with $\Psi^s_{\alpha_i} = (\mathcal{T}_{\alpha_i}, \{U^1_{\alpha_i}, \ldots, U^{s-1}_{\alpha_i}, U^s_{\alpha_i}\})$. Let $Q^s_{\alpha_i}$ be the query of α_i at state s, and $M^s_{\alpha_i}$ the set of preferred abductive stable models of α_i at state s each of which with hypothesis any abductive explanation $\Delta^s_{\alpha_i}$ for $Q^s_{\alpha_i}$, for $1 \leq i \leq n$. Then, $\mathcal{S}^s = \{M^s_{\alpha_1}, \ldots, M^s_{\alpha_n}\}$.

The semantics of the multi-agent system depends on the queries of the agents. In fact, the models of an agent at a certain state depend on the abducibles assumed at that state which in turn depend on the queries of the agent.

6 Preferring Abducibles

In our framework we defined the priority relation over generalized rules. A possible question is: Can we also express preferences over abducibles ? Being able to do so will allow us to perform for example *exploratory data analysis* which aims at suggesting a pattern for further inquiry, and contributes to the conceptual and qualitative understanding of a phenomenon.

Preferences over alternative abducibles can be coded into cycles over negation, and preferring a rule will break the cycle in favour of one abducible or another. Let $\lhd$ be a binary predicate symbol whose set of constants includes all the abducibles in $\mathcal{A}$. $a \lhd b$ means that abducible a is preferred to abducible b.

Assume that a unexpected phenomenon, x, is observed by an agent α, and that α has three possible hypotheses (abducibles), a, b and c, that are capable of explaining it. In exploratory data analysis, after observing some new facts, we abduce explanations and explore them to check predicted values against observations. Though there may be more than one convincing explanation, we abduce only the more plausible ones. Suppose explanations a and b are both more plausible than (and therefore preferable to) explanation c. This can be expressed as:

$$P = \left\{ \begin{array}{l} x \leftarrow a \\ x \leftarrow b \\ x \leftarrow c \end{array} \right\} \text{ and the preferences as: } \left\{ \begin{array}{l} a \lhd c \\ b \lhd c \end{array} \right\}.$$

The program above can be represented by a prioritized abductive logic program $(P', \mathcal{A})$, where $\mathcal{A} = \{a, b, c, a^*, b^*, c^*\}$ and

$$P' = P \cup \left\{ \begin{array}{ll} a \leftarrow a^* \wedge not\, b \wedge not\, c & (1) \\ b \leftarrow b^* \wedge not\, a \wedge not\, c & (2) \\ c \leftarrow c^* \wedge not\, a \wedge not\, b & (3) \\ 1 < 3 & \\ 2 < 3 & \end{array} \right\} \cup \left\{ \begin{array}{l} a^+ \leftarrow not\, a^* \wedge not\, b^+ \wedge not\, c^+ \\ b^+ \leftarrow not\, b^* \wedge not\, a^+ \wedge not\, c^+ \\ c^+ \leftarrow not\, c^* \wedge not\, a^+ \wedge not\, b^+ \\ not\, a \leftarrow a^+ \\ not\, b \leftarrow b^+ \\ not\, c \leftarrow c^+ \end{array} \right\}.$$

The first three rules (called (1), (2) and (3)) code the alternative abducibles a, b and c into cycles over negation, and the following two priority rules state that the first two rules (i.e., rules (1) and (2)) are preferable to the third rule (i.e., rule (3)). Since it is possible in our framework to abduce also negative abducibles, it is needed to add the rules stating that the negative abducibles $not\ a$, $not\ b$ and $not\ c$ are alternative abducibles. This is achieved by the last six rules by introducing the new atoms a^+, b^+ and c^+.

For simplicity, assume that the initial theory of α is $\mathcal{T} = (P', \mathcal{A}, \{\})$ and that there are no updates in the sequence $\mathcal{U}$ of updating programs of α. Then, α has two preferred abductive stable models to explain its observation x: $M_1 = \{a, x, 1{<}3, 2{<}3, a^*, b^+, c^+\}$ and $M_2 = \{b, x, 1{<}3, 2{<}3, b^*, a^+, c^+\}$. By removing the priority atoms, the abducibles superscripted by * and the atoms superscripted by $^+$ from M_1 and M_2, we obtain the intended models of the initial program P, where abducibles a and b are both preferred over c. This technique can be generalized to prefer among sets of abducibles.

In this example, we have only a partial priority theory over abducibles. Thus, we cannot select exactly one abducible (i.e., one model), as it were the case had we a complete priority relation over all abducibles in $\mathcal{A}$. To prefer between a and b, α can perform some experiment *exp* to obtain confirmation (by observing the environment) about thé most plausible hypothesis. For example, by using the following rules (where

e plays the role of the environment):

$$\left\{\begin{array}{l} choose \leftarrow a \\ choose \leftarrow b \end{array}\right\} \text{ together with } \left\{\begin{array}{l} a \Rightarrow \alpha : chosen \\ b \Rightarrow \alpha : chosen \\ choose \Rightarrow \alpha : (not\ chosen \Rightarrow e : exp) \end{array}\right\}.$$

Suppose that agent α has two hypotheses, a and b, that are capable to explain the observed phenomena x, and that α must discover the correct one. α chooses an hypothesis if a or b or both holds:

$$choose \leftarrow a$$
$$choose \leftarrow b.$$

With this knowledge, agent α has still two preferred abductive stable models: $M_1 = \{a, x, choose, \ldots\}$ and $M_2 = \{b, x, choose, \ldots\}$. As *choose* holds in both models, the last active rule is triggerable. When triggered, it will add (at the next state) the active rule $not\ chosen \Rightarrow e : exp$ to the theory of α, and if $not\ chosen$ holds, α will perform the experiment *exp*. The first two active rules are needed to prevent α by performing *exp* when α has chosen one of the abducibles.

7 Acting and Planning

A typical use of abduction is in generating plans in a planning problem (cf. [3,10,13]). Suppose the situation where some information of the planning problem is missing or changes during the planning process. To tackle this situation we must be able to make a (possibly partial) plan, to start executing it, and in case something goes wrong, to detect the anomaly and to replan. This entails the ability to detect such cases and to replan in order to achieve the original goal.

This approach to planning may involve also the ability to generate revisable plans, that is, plans that are generated in stages. After having executed the actions of each stage, we test whether the performed actions have been successful in order to move to the next stage. If some action failed, then we must revise our plan. A successful plan is a revisable plan that has been completely executed such that the obtained final state satisfies the original goal.

To express revisable plans we must be able to express consecutive actions. We represent actions as abducibles. Let $\diamond$ be a binary predicate symbol whose set of constants includes all the abducibles in $\mathcal{A}$. $a \diamond b$ means that action a must be executed before action b.

Typically, for every action a there will be one or more active rules where the action occurs in the body. For example, $a \wedge L_1 \wedge \ldots \wedge L_k \Rightarrow Z$ represents an action whose preconditions are $L_1 \wedge \ldots \wedge L_k$ and whose effect is the project Z.

Suppose that in order to solve a plan p we must execute two actions, a and b, sequentially, a before b. This can be expressed as:

$$\left\{\begin{array}{l} p \leftarrow a \diamond b \\ a \wedge precA \Rightarrow Z_1 \\ b \wedge precB \Rightarrow Z_2 \end{array}\right\}.$$

The program above can be represented by a prioritized abductive logic program $(P, \mathcal{A})$, where $P = \{p \leftarrow a \wedge seq(a,b) \wedge seq(b, true)\}$ and $\mathcal{A} = \{a, b\}$, together with the active rules:

$$R = \left\{ \begin{array}{l} a \wedge precA \wedge seq(a,G) \Rightarrow Z_1 \wedge \alpha{:}(?{-}G) \\ b \wedge precB \wedge seq(b,G) \Rightarrow Z_2 \wedge \alpha{:}(?{-}G) \end{array} \right\}.$$

Suppose now that we want to test whether the action a has been successful before executing b, that is, $p \leftarrow a \diamond test \diamond b$. In case a has been successful, we execute b, otherwise a is executed again. Let *propA* be the property that must hold whenever a has been successful. This can be formalized as: $\mathcal{A} = \{a, b, test\}$, $P = \{p \leftarrow a, seq(a, test), seq(test, b), seq(b, true)\}$ together with the active rules:

$$R = \left\{ \begin{array}{l} a \wedge precA \wedge seq(a,G) \Rightarrow Z_1 \wedge \alpha{:}(?{-}G) \\ test \wedge not\ propA \Rightarrow \alpha{:}(?{-}a) \\ test \wedge propA \wedge seq(test,G) \Rightarrow \alpha{:}(?{-}G) \\ b \wedge precB \wedge seq(b,G) \Rightarrow Z_2 \wedge \alpha{:}(?{-}G) \end{array} \right\}.$$

8 Conclusion and Future Work

We have presented a logical framework of a multi-agent system in which each agent can communicate with and update other agents, can react to the environment, is able to prefer and to abduce hypotheses in order to explain observations made, which in turn may trigger reactions.

Applications in which our agent technology can have a significant potential to contribute, and that we are considering, are internet applications, e.g., information integration and web-site management. To this aim we have designed an agent architecture [14] and we are going to start implementing it. The problem of information integration consists in how to integrate data from multiple heterogeneous sources. The goal of a data integration system is to provide a uniform interface to a several data sources. An example of information integration is the task of providing information about movies from data sources on the web (for example, cf. [15]).

The goal of web site management applications is the flexible construction and modification of web sites. Web sites contain and integrate several pieces of data that are linked together into a navigational structure. One possible approach in these applications is to declaratively represent web sites, that is, to logically represent the kind of data and the structural aspects of web sites. One advantage would be the ability to automatically reconstruct and to enforce integrity constraints on web sites. The underlying idea is that such sites restructure themselves depending on the usage patterns, and can adapt themselves according to the profile of individual users. If we represent declaratively a web site, then we can reason about the structure of the site, for example, to see if the integrity constraints hold any time the site is updated (e.g., no dangling pointers, an employee's homepage should point to the department's homepage, etc.).

We believe that the theory of the agents considered is a rich evolvable basis, and suitable for engineering configurable, dynamic, self-organizing and self-evolving agent societies. In [12, p. 289] Jennings argues that open, networked systems are characterized by the fact that there is no simple controlling organization and that there is constant change. Thus, these systems are collections of independently developed software entities that are interacting with each other. Jennings claims that the computational model of these systems places several requirements:

(i) the individual entities need to be able to act to achieve specified objectives (i.e., they must be active and autonomous);
(ii) the individual entities need to be reactive and proactive;
(iii) the computational entities need to be capable of interacting with entities that were not foreseen at design time (i.e., they must be able to engage in flexible interactions);
(iv) any organizational relationships that do exist must be reflected in the behaviour and actions of the agents (i.e., the organizational relationships must be explicitly represented).

Our agents have those abilities. Regarding to (iv) we can represent dynamic organizational structures in the theory of agents that can self-organize themselves and self-evolve via the ability of preferring and updating.

Davidsson [8] categorises agent societies based on the following properties:

- *openness*, i.e., the possibility for agents to join the society without any restriction;
- *flexibility*, i.e., the degree to which agents are restricted in their behaviour by the society;
- *stability*, i.e., predictability of the consequences of actions, and
- *trustfulness*, i.e., the extent to which the owners of the agents trust the society.

Davidsson argues that whereas open societies support openness and flexibility, closed societies support stability and trustfulness. In several systems, e.g., in systems characterized as information eco-systems, there is a need for societies that support all these aspects. Thus, he investigates other kinds of agent societies among which *semi-open societies*, where any agent can join the society provided that it follows some well-specified restrictions.

The agents that we have presented can be used for example to animate semi-open societies. In fact, to do so, the agents must exhibit several rational abilities, e.g., they must be able to decide which society to join according to their own goals, to decide whether or not to accept the restrictions imposed to enter into the society, and eventually to negotiate them. Thus, to increase the degree of trustfulness and stability in open societies we need to empower the agent theories with rational abilities to make the agents able to react to events unforeseen at design time.

Another framework where our agents can be of use is the one proposed by Vercouter [16]. He proposes a distributed approach to handle openness in multi-agent systems. Typically, the openness of a system is ensured by an entity, called the middle-agent, that supports the addition and removal of agents from the system, and knows the capabilities of all the agents. The middle-agent handles also the connection problem: when an agent looks for another agent that provides a service it needs. The middle-agent (centralized) approaches to the problem of openness have a number of limits, and to overcome them Vercouter investigates a distributed approach (i.e., without the middle-agent technique) based on the notion of *presentation/recommendation protocol*. Each agent can have its own representation of other agents, and can detect its possibilities of cooperation. When an agent joins a coalition, the agent presents itself to some other agents of the coalition and informs them of its capabilities. During this phase, the agent can ask them for recommendations of other agents that provide services that it needs, and start a presentation protocol with them. This is clearly another framework where several rational abilities are needed in agents, for example, to reason about, to evaluate and to prefer which agents are relevant for its purposes.

Within the proposed multi-agent system framework we can represent groups, teams, coalitions of agents implicitly based on the internal mental states of the members. It is advocated, especially in open multi-agent systems (cf. [5,12,17]), that there is a need to make the organisational elements as well as the formalisation of the agent interactions of a multi-agent system externally visible rather than being embedded in the mental state of each agent, i.e., it is needed to explicitly represent the organisational structure and the agent interactions. For example, we may formalize a group structure as an abstract description of a group that identifies all the roles and interactions that can occur within a group. This will allow us to precisely define concepts like permission, obligation, responsibility, social laws, requirements and roles of the society, etc. We are investigating how to explicitly represent organisational structures in our framework, and how to animate them with agents in a way that each agent will automatically have a view (perhaps partial) of the organisational structure and the externally visible events. We believe that this can be achieved through the concept of *organisational reflection*[3]. In fact, we may embed the theory of agents with various degrees of reflection abilities to make an agent able to introspect various aspects of the organisation where it belongs. We may also imagine a more complex organisational structure composed of several nested organisational substructures each of which with its own specific rules and with more general rules inherited by the parent structure. In this way, an agent may only have a partial view of the entire organisational structure. Thus, the members of a society will have information of it (e.g., they will know its rules, requirements and laws), and they will have more information when reasoning about other agents' actions. In addition, the agents will be able to reason upon organisational structures and eventually try to modify them. In this way organisational structures will not be rigid, but flexible and can evolve with the agents' intervention.

Organisational structures and rules are needed to provide control over the member agents with respect to their actions and interactions.

Castelfranchi [7] argues that social order is a major problem in multi-agent systems. The idea of a total control and a technical prevention against chaos, conflicts and deception in computers is unrealistic and even self-defeating in some case, like in the building trust. He states that there is some illusion in Computer Science about solving this problem by rigid formalization and rules, constraining infra-structures, security device, etc. and there is skepticism or irritation towards more soft and social approaches, that leave more room to spontaneous emergence, or to decentralized control, or to normative "stuff" which is not externally imposed but internally managed by the agents. He claims that the most effective solution to this problem is social modelling and that

[3] This term was first introduced by J. Ferber and O. Gutknecht [11], but it is used here with a different meaning.

it should leave some flexibility and try to deal with emergent and spontaneous form of organizations (that is, decentralized and autonomous social control). The problem with this approach is that of modeling the feedback from the global results to the local/individual layer.

To solve this problem, we need two ingredients (still missing in our agent theory): the ability of introspection and meta-reasoning. In fact, in order to be able to dynamically change the organization, structure of the multi-agent system, agents must be aware (even if partially) of the structure and must be able to introspect about it. By using meta-reasoning the agent can evaluate it, obtain feedback from it and eventually try to modify it via preferences and updates in a rational way. Our ongoing work and future projects is discussed in [1].

In conclusion, our claim is that in order to have dynamic, flexible agent societies we need to have suitable agent theories, otherwise the structure modeling the agent society will be rigid in the sense that it will not be modifiable by the agents themselves.

Acknowledgements

L. M. Pereira acknowledges the support of PRAXIS project MENTAL and POCTI project FLUX.

References

1. J. J. Alferes, P. Dell'Acqua, E. Lamma, J. A. Leite, L. M. Pereira, and F. Riguzzi. A logic based approach to multi-agent systems. ALP Newsletter, August 2001. Available at `http://centria.di.fct.unl.pt/~lmp`.
2. J. J. Alferes, J. A. Leite, L. M. Pereira, H. Przymusinski, and T. C. Przymusinski. Dynamic updates of non-monotonic knowledge bases. *J. Logic Programming*, 45(1-3):43–70, 2000.
3. J. J. Alferes, J. A. Leite, L. M. Pereira, and P. Quaresma. Planning as abductive updating. In D. Kitchin, editor, *Procs. of the AISB'00 Symposium on AI Planning and Intelligent Agents*, pages 1–8, Birmingham, England, 2000. AISB.
4. J. J. Alferes and L. M. Pereira. Updates plus preferences. In M. O. Aciego, I. P. de Guzmn, G. Brewka, and L. M. Pereira, editors, *Logics in AI, Procs. JELIA'00*, LNAI 1919, pages 345–360, Berlin, 2000. Springer.
5. A. Artikis and G. Pitt. A formal model of open agent societies. Proc. of Autonomous Agents, 2001.
6. G. Brewka and T. Eiter. Preferred answer sets for extended logic programs. *Artificial Intelligence*, 109:297–356, 1999.
7. C. Castelfranchi. Engineering Social Order. In Andrea Omicini, Robert Tolksdorf, and Franco Zambonelli, editors, *Engineering Societies in the Agents World. 1st Int. Workshop ESAW 2000. Revised Papers*, LNAI 1972, pages 1–18, Berlin, 2000. Springer-Verlag.
8. P. Davidsson. Categories of artificial societies. In A. Omicini, P. Petta, and R. Tolksdorf, editors, *ESAW 2001*, pages 1–9, 2001. Available at: `http://lia.deis.unibo.it/confs/ESAW01/`.
9. P. Dell'Acqua and L. M. Pereira. Updating agents. In S. Rochefort, F. Sadri and F. Toni (eds.), Procs. of the ICLP'99 Workshop on Multi-Agent Systems in Logic (MASL'99), 1999.
10. M. Denecker and A. Kakas, editors. *Special Issue on Abductive Logic Programming*, volume 44(1-3). J. Logic Programming, 2000.

11. J. Ferber and O. Gutnecht. A meta-model for the analysis and design of organisations in multi-agent systems. In *Int. Conf. on Multi-Agent Systems (ICMAS-98)*, pages 128–135. IEEE Computer Society, 1998.
12. N. R. Jennings. On agent-based software engineering. *Artificial Intelligence*, 117:277–296, 2000.
13. A. C. Kakas, R. S Miller, and F. Toni. Planning with incomplete information. In C. Baral and M. Truszczynski, editors, *Proc. NMR'2000, 8th Int. Workshop on Non-Monotonic Reasoning, Spec. Session on Representing Actions and Planning*, 2000.
14. J. A. Leite, J. J. Alferes, and L. M. Pereira. Minerva - A Dynamic Logic Programming Agent Architecture. To appear in: ATAL01 - 8th Int. Workshop on Agent Theories, Architectures, and Languages, 2001.
15. A. Y. Levy and D. S. Weld. Intelligent internet systems. *Artificial Intelligence*, 118:1–14, 2000.
16. L. Vercouter. A distributed approach to design open multi-agent systems. In A. Omicini, P. Petta, and R. Tolksdorf, editors, *ESAW 2001*, pages 24–37, 2001. Available at: `http://lia.deis.unibo.it/confs/ESAW01/`.
17. F. Zambonelli, N. R. Jennings, and M. Wooldridge. Organisational abstractions for the analysis and design of multi-agent systems. In P. Ciancarini and M. Wooldridge, editors, *Agent-Oriented Software Engineering*, LNCS 1957, pages 127–141, Berlin, 2001. Springer-Verlag.

Reasoning about Failure

László Aszalós* and Andreas Herzig

IRIT, Universite Paul Sabatier
118 route de Narbonne, F-31062 Toulouse Cedex 4
{aszalos,herzig}@irit.fr

Abstract. In this paper we investigate a modal logic of believing and saying to reason about unreliable agents in a system of communicating agents. We suppose that communication is reliable and semi-public: an agent's utterances are communicated to all the adjacent agents. We suppose that to each agent is associated some set of facts that he monitors, and that his perception is perfect in what concerns these facts. We show how an agent can detect failure of another agent by deduction in our logic. To that end we use a tableau theorem prover for our logic.

1 Introduction

Often in the literature the agents and their communication are ideal, but in real life this doesn't hold. We can use error-correcting codes to eliminate errors due to the communication channel, but according to Murphy's Law the agents may have a breakdown and get unreliable. In this case they may communicate false statements. (Nevertheless, it is not necessarily the case that every statement they utter is false.) This suggests a link with logics of truthtellers and falsetellers (or knights and knaves), as used in formalization of Smullyan's puzzles. Being reliable corresponds to being a truthteller in the puzzles. The present approach is derived from work on such logics [1].

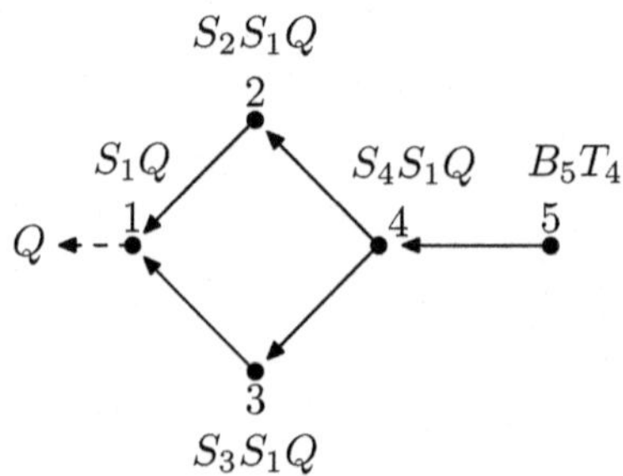

Fig. 1. The Communication Lines between the Agents.

Consider a toy-example with five persons (agents). Not all the agents can communicate with each other, only the agents that are connected as specified in Fig. 1. Note that

* On leave from University of Debrecen, Hungary, aszalos@math.klte.hu.

A. Omicini, P. Petta, and R. Tolksdorf (Eds.): ESAW 2001, LNAI 2203, pp. 74–85, 2001.

this is a reasonable assumption in particular in large networks, where public communication is too costly. Note also that in our example the communication isn't symmetric, but we don't exclude the symmetric case. Person 1 is a banker, and all the others know it. Only person 1 knows that his bank is getting bankrupt (Q). He informed person 2 and 3 (S_1Q) about it. They (independently) reported this to person 4 (S_2S_1Q and S_3S_1Q). Person 4 doesn't trust neither of them, but if each states it, she will believe that person 1 really said this. Person 5 has a big amount of money in this bank, so person 4 inform person 5 about this bankrupt (S_4S_1Q) before this information became public and the bank closes all the accounts. Person 5 trusts in person 4, so she will believe in bankrupt and take out her money.

Let $\mathcal{AGT} = \{i, j, ...\}$ be the set of the agents. The propositional letter T_i denotes that agent i is reliable, for $i \in \mathcal{AGT}$.

We associate a modal operator S_i to every $i \in \mathcal{AGT}$. The formula S_iA is read "agent i said A". The modal operator S_i is non-normal [4], and is hence neither closed under logical truth, nor logical consequence, conjunction, material implication [2]. We shall even suppose that it is not closed under logical equivalence.

The relation between reliability and the modal operator of saying is the following: if a reliable agent said something, then that is true. This is axiomatized by the axiom schema:

$$T_i \rightarrow (S_iA \rightarrow A) \tag{1}$$

The formula $T_i \rightarrow (A \rightarrow S_iA)$ isn't valid: infinitely many true statements exist, and the reliable agents don't need to announce all of them. For example the communication is costly, so the agents don't report all the facts they know, only a small subset of it. Note that unreliable agents do not always lie, i.e. $\neg T_i \rightarrow (S_iA \rightarrow \neg A)$ should not be valid. This contrast with the logics of truthtellers and falsetellers as used in formalization of Smullyan's puzzles, where falsetellers never tell true statements.

We associate a modal operator of belief B_i to every $i \in \mathcal{AGT}$. The formula B_iA is read "agent i believes that A". We adopt the modal logic KD45 as the logic of belief. This implies that we suppose that agents cannot entertain inconsistent beliefs, and that they are aware of their beliefs and of their disbeliefs.

We consider that agents might be aware of their own unreliability, but are not necessarily so. Hence there is no formal link between reliability and the modal operator of belief, i.e. we neither have B_iT_i nor $\neg T_i \rightarrow B_i\neg T_i$. Hence it might be the case that an agent believes to be unreliable but is not, and the other way round that he believes to be reliable but is not.

We describe with propositional letters the facts of the world of the agents. An agent cannot monitor all the facts, for example a web-agent cannot read all the web pages, a satellite cannot see the whole surface of the Earth. Hence agent i can monitor only a subset of facts. In our example, $Q \in obs(1)$. We suppose that the set $obs(i)$ is a set of atoms. Note that it is possible that $T_i \in obs(j)$, that is agent j monitors agent i, so agent j always knows that agent i is reliable or not. In the special case, where i and j are the same, then $T_i \in obj(i)$, that is agent i can monitor himself, so i knows whether he is reliable or not. The former case can be realistic in some electronic circuit, but the latter (at least in our opinion) not. If $A \in obs(i)$, then i knows (believes) whether the fact is true or not:

$$A \rightarrow B_i A \qquad \text{if } A \in obs(i) \tag{2}$$

and

$$\neg A \rightarrow B_i \neg A \qquad \text{if } A \in obs(i) \tag{3}$$

We remark that $obs(i)$ can be defined as a set of arbitrary formulae.

What is the relation between (2) and (3) and the notion of competence? By definition i is competent at A if $B_i A \rightarrow A$ and $B_i \neg A \rightarrow \neg A$ hold. It is easy to check that if $A \in obs(i)$ then by (2) and (3) i is competent at A. Note that there is no formal link between truthtelling and competence. It the doctor knows that the bad news will shorten the patient's life, then he won't utter that the patient is incurable. He need to say something so he will lie. In this case the doctor is competent at disease but "unreliable".

In this article we shall examine only the simplest case. This is the reason why we work only with the modal operators 'believes' and 'said'. In the axioms before we could use modal operator 'knows' instead of 'believes' because if somebody monitors something then he exactly knows what happened (and not only believes in it). Instead of using a new modal operator, we suppose here that $K_i A$ is defined as $B_i A \wedge A$. We suppose that the monitored facts are common knowledge. In particular we have $B_k(A \rightarrow B_i A)$ and $B_k(\neg A \rightarrow B_i \neg A)$ whenever we have $A \in obs(i)$.

In multi-agent communication there is public and point-to-point communication. If there are many agents, then public communication is costly. If there is only point-to-point communication then it is hard to detect malfunction. For these reasons we shall work with a semi-public communication method: if some agent announces something, then all the nearby agents perceive it. We introduce a reflexive, non-transitive relation $Listen$. $Listen(i, j)$ means that the agent i listens agent j's reports. In this case agent j knows (believes) what agent i reported about formula A or not:

$$S_j A \rightarrow B_i S_j A \qquad \text{if } Listen(i, j) \tag{4}$$

and

$$\neg S_j A \rightarrow B_i \neg S_j A \qquad \text{if } Listen(i, j) \tag{5}$$

Note that as the modal operator B_j is KD45, the converse of the above implications can be proved, too. Note also that $S_j A \rightarrow B_j S_j A$ and $\neg S_j A \rightarrow B_j \neg S_j A$, because $Listen$ is reflexive.

We suppose that the communication lines are common knowledge. In particular we have $B_k(S_j A \rightarrow B_i S_j A)$ and $B_k(\neg S_j A \rightarrow B_i \neg S_j A)$ whenever we have $Listen(i, j)$.

2 Semantics

The modal operator of belief is a normal modal operator, and its possible worlds semantics is based on accessibility relations. Let us note R_i the accessibility relation associated to modal operator B_i.

The modal operators of saying are non-normal. Similar to Fagin et al.'s [6] awareness modality, we associate a subset of formulae $\vartheta_i(w)$ to every $i \in \mathcal{AGT}$ and possible world w, and stipulate:

$$w \models S_i A \text{ iff } A \in \vartheta_i(w).$$

Hence our models are of the form $\langle W, \langle R_i\rangle_{i\in\mathcal{AGT}}, \langle\vartheta_i\rangle_{i\in\mathcal{AGT}}, I\rangle$, where I associates to every world an interpretation I_w mapping propositional letters to truth values. Every R_i is transitive, serial and euclidian. Moreover models must satisfy some constraints that we introduce in the sequel.

A reliable agent tells the truth. The other way round, if an agent said some true statement, then from this we cannot abduce that he is reliable. For example if an agent said that he is unreliable: $S_i\neg T_i$, then this statement is true and he isn't reliable. To avoid the Liar Paradox we need to define semantics of T_i carefully: we treat T_i as a propositional letter and enable only the models where the following constraint holds:

$$\text{if } I_w(T_i) = 1 \text{ then for all } A \in \vartheta_i(w), w \models A$$

This ensures validity of Axiom 1.

The agent i monitors the facts in $obs(i)$, hence i knows exactly which are true and which are not, so the facts in $obs(i)$ have the same truth value in the actual world and in worlds accessible for i:

$$I_w|_{obs(i)} = I_{w'}|_{obs(i)} \text{ if } w'R_iw$$

This restriction makes Axiom 2 and 3 valid.

If agent i listens to what agent j reports then agent i has correct beliefs about agent j's reports, so what agent j said is the same in the actual world and the worlds accessible for i.

$$\text{If } Listen(i,j) \text{ and } wR_iw' \text{ then } \vartheta_j(w) = \vartheta_j(w')$$

This restriction makes Axiom 4 and 5 valid.

3 Reasoning about Communication and Failure

We can prove easily the following theorems.

Theorem 1.

$$B_iT_i \wedge S_iA \rightarrow B_iA.$$

Proof. $T_i \wedge S_iA \rightarrow A$ is an axiom. Then $B_iT_i \wedge B_iS_iA \rightarrow B_iA$ by principles of modal logic K. As we have stipulated that $Listen$ is reflexive, we have $S_iA \rightarrow B_iS_iA$ by axiom (4). Hence $B_iT_i \wedge S_iA \rightarrow B_iA$ □

If moreover $T_i \in obs(i)$ then we get:

Theorem 2. *If $T_i \in obs(i)$ then*

$$T_i \wedge S_iA \rightarrow B_iA.$$

If agent k got contradictory information from agent i and j, then one of them isn't reliable:

Theorem 3. *If $Listen(k, i)$ and $Listen(k, j)$ then*

$$S_iA \wedge S_j\neg A \rightarrow B_k(\neg T_i \vee \neg T_j).$$

If agent k believes that agent i or agent j is reliable, and agent i and j report the same, then agent k will believe in this information:

Theorem 4. *If $Listen(k, i)$ and $Listen(k, j)$ then*

$$(S_iA \wedge S_jA \wedge B_k(T_i \vee T_j)) \rightarrow B_kA.$$

If agent k is careful, he can wait until n agents report the same information before he believes in it.

Theorem 5. *If $Listen(k, i_j)$ $j = 1, \ldots, n$ and $i_j \neq i_l$ then*

$$(S_{i_1}A \wedge \cdots \wedge S_{i_n}A \wedge B_k(T_{i_i} \vee \cdots \vee T_{i_n})) \rightarrow B_kA.$$

More generally, agent k may suppose that there are at most $n - 1$ unreliable agents.

Theorem 6. *If $Listen(k, i_j)$ $j = 1, \ldots, n$, $i_j \neq i_l$ and* $B_k\left(\bigwedge_{i_j \neq i_l}(T_{i_1} \vee \cdots \vee T_{i_n})\right)$ *then*

$$(S_{i_1}A \wedge \cdots \wedge S_{i_n}A) \rightarrow B_kA.$$

This illustrates that our logic is flexible enough to allow agents to hold various kinds of beliefs about reliability: we can express confident attitudes, such as that of an agent taking over a communicated belief as soon as some threshold is attained (e.g. that of at least two agents saying the same thing like in our example). We can as well express suspicious attitudes of waiting until all or all but one adjacent agents have communicated the same belief. Finally, agents may take a more balanced attitude and take over a communicated belief A if the number of adjacent agents having communicated A is greater than the number of adjacent agents having communicated $\neg A$ (disregarding those adjacent agents who didn't say anything about A).

Using the hypothesis that the agents know the communication lines, we obtain the following theorem.

Theorem 7. *If $Listen(k, i)$ and $Listen(j, i)$ then*

$$S_iA \rightarrow B_kB_jS_iA.$$

Proof. The steps are the following:

1)	S_iA	hypothesis
2)	B_kS_iA	from 1 by Axiom 4 and $Listen(k, i)$
3)	$B_k(S_iA \rightarrow B_jS_iA)$	agents know the axioms
4)	$B_kB_jS_iA$	from 2 and 3

□

Finally, using our assumption that the monitored facts are known by all the agents, we obtain:

Theorem 8. *If $A \in obs(i) \cap obs(j)$ then*

$$A \rightarrow B_iB_jA \text{ and } A \rightarrow B_jB_iA.$$

4 Tableaux

The Lotrec generic tableau prover [3] offers an easy way to implement a sound and complete tableau method for our logic. The rules of the classical logic connectives and of belief are the usual ones. If a node contains an unmarked formula S_iA then we add to this node the formula $T_i \rightarrow A$ and mark S_iA, not to use this rule again (Fig. 2). Note if a node contains formula $\neg S_iA$ then of course we cannot "open" this formula: if

Fig. 2. Lotrec Rule for Saying.

somebody did not say A, we cannot abduce anything (except just that). In other words there is no formula that is inconsistent with $\neg S_iA$ except S_iA.

If a tableau node n_2 is accessible from n_1 via R_i and i listens to j then we can copy all formulae of the form S_jA and $\neg S_jA$ from node n_2 to n_1 and vice versa (Fig.3 and 4). Fig. 5 shows that Axiom 4 can be proved with this rule.

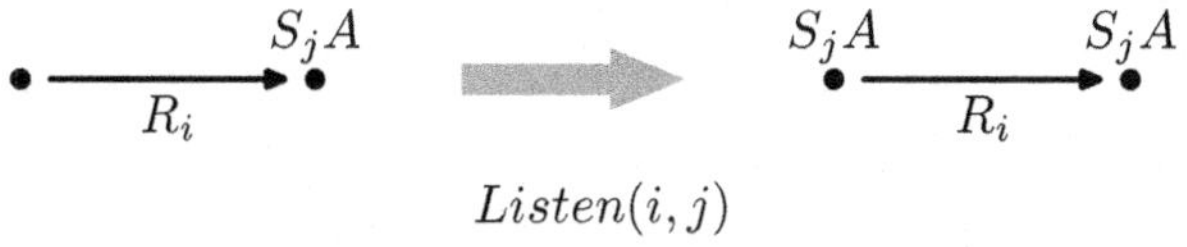

Fig. 3. Lotrec Rule for $Listen(i, j)$.

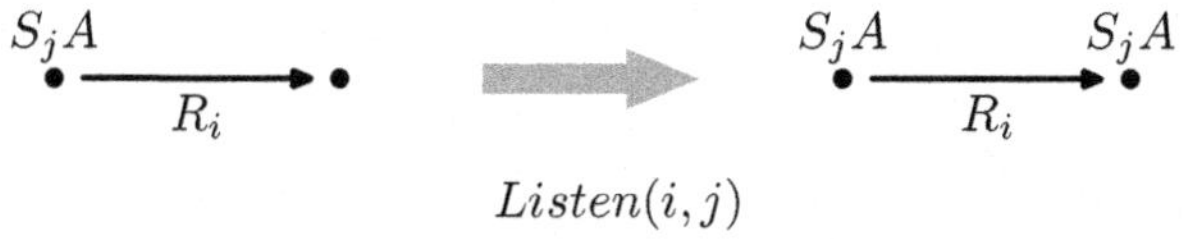

Fig. 4. Lotrec Rule for $Listen(i, j)$.

Similarly if the node contains a propositional letter $A \in obs(i)$, and the node is accessible from another node by R_i then we copy this propositional letter to that node (Fig. 6). The other way round, if a node contains $A \in obs(i)$, and another node is accessible from this node, then we copy A to that node (Fig. 7).

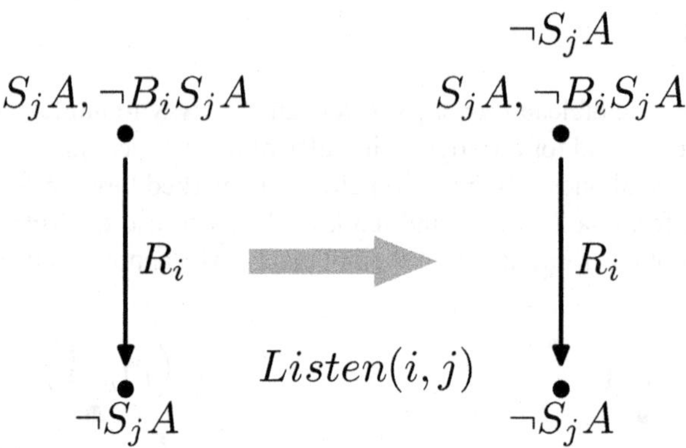

Fig. 5. Proof of Axiom 4.

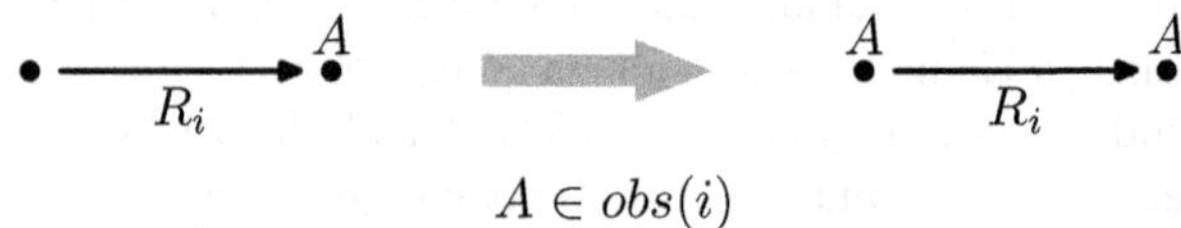

Fig. 6. `Lotrec` Rule for $obs(i)$.

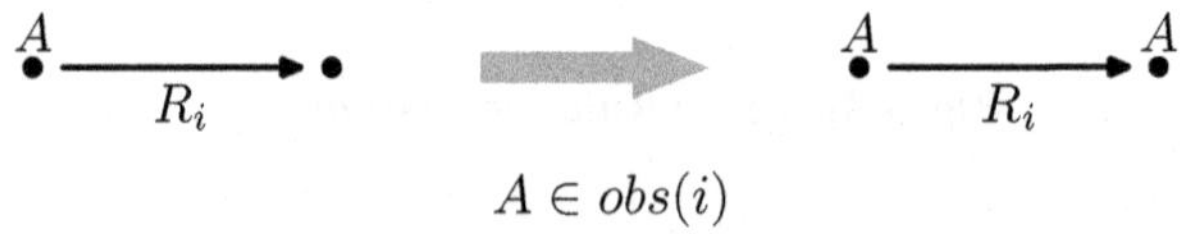

A A A

R_i R_i

$A \in obs(i)$

Fig. 7. `Lotrec` Rule for $obs(i)$.

5 Example in Detail

In our example the hypotheses are Q, S_1Q, S_2S_1Q, S_3S_1Q, $B_4(S_2A \wedge S_3A \rightarrow A)$. The last formula expresses that if the utterances of person 2 and 3 agree then person 4 adopts them. We moreover have S_4S_1Q, B_5T_4, $Q \in obs(1)$ and $Listen(i, j)$ if agent i and j are connected as specified in Fig. 1. We want to prove that B_1Q, B_2S_1Q, B_3S_1Q, B_4S_1Q and B_5Q.

1)	Q	hypothesis
2)	B_1Q	1 and Axiom 2
3)	S_1Q	hypothesis
4)	B_2S_1Q	3, $Listen(2,1)$ and Axiom 4
5)	S_2S_1Q	hypothesis
6)	$B_4S_2S_1Q$	5, $Listen(4,2)$ and Axiom 4
7)	B_3S_1Q	3, $Listen(3,1)$ and Axiom 4
8)	S_3S_1Q	hypothesis
9)	$B_4S_3S_1Q$	8, $Listen(4,3)$ and Axiom 4
10)	$B_4(S_2S_1Q \wedge S_3S_1Q \rightarrow S_1Q)$	hypothesis, where A is S_1Q
11)	$B_4S_2S_1Q \wedge B_4S_3S_1Q \rightarrow B_4S_1Q$	10
12)	B_4S_1Q	6, 9 and 11
13)	S_4S_1Q	hypothesis
14)	$B_5S_4S_1Q$	13, $Listen(5,4)$ and Axiom 4
15)	B_5T_4	hypothesis
16)	B_5S_1Q	14, 15
17)	B_5Q	16, Axiom 2 and 3

Note that we could have given a tableau proof as well.

If the hypotheses are the same as before but $\neg B_5T_4$, we cannot prove formula B_5S_1, only $B_5S_4S_1Q$. It is interesting, that if the conditions are the same but the hypothesis are only S_4S_1Q and B_5T_4 then we can prove that B_5S_1Q, even if $\neg S_1Q$.

In the title of the paper we promised reasoning about failure, so let us see how can we use our logic to detect failure. The communication lines are the same as before, and person 5 observes Q, too. The hypotheses are the following:

Q	bankrupt
S_1Q	person 1 informs person 2 and 3
S_2S_1Q	person 2 informs person 4
$S_3S_1\neg Q$	person 3 lies

By Theorem 3 person 4 knows that one of them is lying, but he doesn't know who and he cannot deduce anything about bankrupt.

$S_4(S_2S_1Q \wedge S_3S_1\neg Q)$ person 4 informs person 5 about statements he got

Person 5 knows that there is a bankrupt. He got only person 4's message, and only from this he cannot abduce that person 3 is unreliable. If person 4 is unreliable then he can say this even if person 3 didn't say anything. If person 5 believes in the reliability of person 4 then he can conclude that person 3 is unreliable.

6 Discussion and Related Work

6.1 Wooldridge, Lomuscio, et al.

In a series of papers, Wooldridge, Lomuscio et al. have investigated logic VSK [10]. Within this logic they study the interplay of notions of visibility, seeing, and knowledge. V_iA means that A is accessible (visible) for agent i, S_iA means that i perceives A, and K_iA means that i knows that A. They present a traffic example involving a faulty robot b refusing to give way to another robot a. The main mechanism to detect b's malfunction is based on a's knowledge that b does not perceive the danger of collision, although it is visible for b: $K_a(V_b\text{coll} \wedge \neg S_b\text{coll})$.

We can express this scenario in our logic as follows: the hypotheses are

- $Listen(a, b)$;
- r-o-w$_a \in obs(a)$, i.e. a observes whether he has right of way;
- $B_a(\text{r-o-w}_a \wedge \neg T_b) \rightarrow \text{ev-act}_a$, i.e. if a believes that he has priority and that b isn't reliable, then a undertakes an evasive action.

We must prove that r-o-w$_a \wedge S_b\neg$r-o-w$_a \rightarrow$ ev-act$_a$, i.e. if a has priority, and b does not give way to a, then a takes an evasive action. This can be established using necessitation on the instance $T_b \wedge S_b\neg\text{r-o-w}_a \rightarrow \neg\text{r-o-w}_a$ of axiom (1), resulting in $B_a(T_b \wedge S_b\neg\text{r-o-w}_a \rightarrow \neg\text{r-o-w}_a)$. Now as r-o-w$_a \in obs(a)$, we get B_ar-o-w$_a$. Then with standard principles of modal logic we obtain $B_a\neg T_b$. Finally, from our last hypothesis we obtain ev-act$_a$.

In comparison, in Wooldridge, Lomuscio et al.'s logic reliability of an agent is not a 'first-class citizen', whereas in our logic it is. Suppose we have to express some general rules of behavior such as a taking some action of warning other agents whenever a believes b is unreliable. This can be done in our logic by $B_a\neg T_b \rightarrow \text{warn-act}_a$. In their logic, this must be written down for every piece of information potentially revealing b's malfunction.

Moreover, in Wooldridge, Lomuscio et al.'s logic there is no communication, but only perception. Therefore, if a fact isn't visible for an agent then he cannot get information about it, and hence an agent cannot get information in an indirect way. We think that such communication features are important in particular in big agent societies.

The expressivity of our logic is highlighted by the fact that we can distinguish three cases of conflict between what is believed and what is said:

- $B_iA \wedge \neg S_iA$
- $B_iA \wedge S_i\neg A$
- $\neg B_iA \wedge S_iA$

6.2 Liau

Liau [9] has proposed a logic of belief, trust and information acquisition. He uses $I_{ij}A$ to express that agent i acquires information A from j, and he uses $T_{ij}A$ to express that agent i trusts agent j's judgment on the truth of A.

In our language we can express $I_{ij}A$ by S_iA, under the hypothesis that j listens to i, i.e. $Listen(i, j)$. Our notion of trust is a more global one than Liau's: our formula B_iT_j means that agent i trusts *all* judgments of agent j. The reason is that we only distinguish between the case where an agent has a breakdown and the case where he is reliable: if i supposes malfunction of j, then i should doubt every formula received from j, while in the case of reliability, i should trust every formula received from j.

In our approach malfunction can be deduced as soon as some inconsistency between agents' statements has been discovered. Liau's trust operator being relativized, one cannot deduce breakdown of an agent in his approach. The other way round his trust operator allows a more fine-grained analysis in many cases. We have two possibilities to express trust in our logic. The first is to define $T_{ij}A$ as $B_i(S_jA \rightarrow A) \wedge B_i(S_j\neg A \rightarrow \neg A)$ and the second is $B_i(B_jA \rightarrow A) \wedge B_i(B_j\neg A \rightarrow \neg A)$.

Another difference is that in Liau's language there is point to point communication. Thus the agents can get direct information. In our case there is semi-public communication, and the agents sometimes get indirect information. In order to decrease the cost of the communication the agents can communicate the summary of the information they got. In our example agent 4 said S_1Q instead of $S_3S_1Q \wedge S_2S_1Q$. Here it is hard to distinguish direct and indirect information, as Liau did.

6.3 Demolombe

Demolombe [5] has proposed a logic of belief, knowledge, and information to reason about trust. He uses a modal operator of belief B_i, and a modal operator of information I_{ij}. Symmetrically to Liau's operator, $I_{ij}A$ expresses that agent i informs agent j about A. Within this logic, which contrarily to Liau's and ours contains no trust operator, Demolombe then analyzes various aspects of the notion of trust. He proposes four basic concepts associated to the interplay between truth, belief and utterances. Agent j trusts i w.r.t. *sincerity* if j knows that i believes what he says:

$$K_j(I_{ij}A \rightarrow B_iA)$$

j trusts i w.r.t. *credibility* if j knows that what i believes is true:

$$K_j(B_iA \rightarrow A)$$

j trusts i w.r.t. *cooperativity* if j knows that i utters his beliefs:

$$K_j(B_iA \rightarrow I_{ij}A)$$

j trusts i w.r.t. *vigilance* if j knows that what is true is believed by i:

$$K_j(A \rightarrow B_iA)$$

Note that, contrarily to us and just as in Liau's case, trust is always about some proposition.

If we replace I_{ij} by S_i and suppose that i listens to j (i.e. we have $Listen(i,j)$), then we can express all these concepts in our logic. As our logic enables us to explicit appropriate hypotheses about reliability and observation, three of them have quite close counterparts that take the form of theorems.

In the case of sincerity, as $B_iT_i \wedge S_iA \rightarrow B_iA$ is provable in our logic (Theorem 1),

$$K_j(B_iT_i \wedge S_iA \rightarrow B_iA)$$

is a theorem, too. We note that our theorem requires a hypothesis slightly stronger than Demolombe's: not only j must believe that i is reliable, but also i himself must do so. This takes into account the (perhaps marginal) case where i ignores his own reliability, and consequently does not believe his own utterances.

Credibility corresponds to our theorem

$$K_j(B_i A \to A) \qquad \text{if } A \in obs(i)$$

which can be derived from axiom 3 and the principle D.

Vigilance corresponds to our theorem

$$K_j(A \to B_i A) \qquad \text{if } A \in obs(i)$$

which can be derived from axiom 2 by necessitation.

Cooperativity has no counterpart. Indeed, we are not able to relativize it to particular formulas, and it is not reasonable to suppose that agents utter just everything they believe. For example the Echelon (spy) system listens a huge mass of messages, but it saves (and utters for its operators) only a little part of it. In our paper the agents have autonomy in their communication: they decide which information they utter and which not. Note that this is independent of cooperativity. We shall consider in further work how this can be integrated, possibly based on a metalinguistic function $publ(i)$ associating to an agent i a set of formulas that are his potential public beliefs, i.e. the beliefs that he utters as soon as he comes to believe them. For example we could add the following axiom:

$$B_i A \to S_i A \qquad \text{if } A \in publ(i).$$

This and the ability of asking could produce a very practical logical language to describe and reason about communication between agents.

6.4 Merging Operations

When an agent takes into account what he learns from other agents, he merges different information sources. Such merging operations have been studied recently [7,8]. Formally, a knowledge set is a multiset $A_1 \sqcup \ldots \sqcup A_n$ containing the different pieces of information $A_1, \ldots, A_n$, and the operator Δ maps a knowledge set $A_1 \sqcup \ldots \sqcup A_n$ to a formula $\Delta(A_1 \sqcup \ldots \sqcup A_n)$.

We identify the elements of a multiset $A_1 \sqcup \ldots \sqcup A_n$ with the utterances of n different agents $S_1 A_1 \wedge \ldots \wedge S_n A_n$, and we identify checking whether $\Delta(A_1 \sqcup \ldots \sqcup A_n) \to A$ with checking the validity of $B_0 H \wedge S_1 A_1 \wedge \ldots \wedge S_n A_n \to B_0 A$, where

- 0 is an agent different from $1, \ldots, n$;
- $Listen(0, i)$ for $0 \leq i \leq n$;
- H is some hypothesis about the number of reliable agents. An example is $\bigwedge_{i \neq j}(T_i \vee T_j)$.

We thus obtain a weak form of merging. Several postulates that every merging operation should satisfy have been put forward in the literature. It turns out that only very few of them hold when translated to our logic. This is in part due to the fact that, just as revision operators, merging operators are part of the metalanguage, and the postulates appeal to consistency, which cannot been expressed within the logic.

7 Conclusion

We have defined a modal logic of believing and saying. We have supposed that utterances are semi-public and that the agents know the communication lines. Under these hypotheses we formalized principles of communication with agents that might fail.

In our approach the adoption of beliefs can be expressed in several ways, for example an agent adopts the information 1) if n different agents said it, 2) if at least one reliable agent said it, 3) if an agent he trust said it. Note that in some domains there is no level of evidence, as Liau said "…a source can only be trusted or distrusted and no intermediary degree is allowed…" For example people believe in God or not: there is no person who believes a bit in him.

Our approach is close to work of Demolombe and Liau, and most of their concepts have a counterpart in our logic. We moreover have studied the theorem proving aspect of our logic, and have given a tableau calculus.

References

1. László Aszalós. The Logic of Knights, Knaves, Normals, and Mutes. *Acta Cybernetica*, 14:533–540, 2000.
2. László Aszalós. Said and Can Say in Puzzles of Knights and Knaves. In *Proc. Premières Journées Francophones sur les Modèles formels pour interaction*, Toulouse, May 2001.
3. Luis Fariñas del Cerro, David Fauthoux, Olivier Gasquet, Andreas Herzig, Dominique Longin, and Fabio Massacci. Lotrec: The Generic Tableau Prover for Modal and Description Logics (system abstract). In *Proc. Int. Joint Conf. on Automated Reasoning (IJCAR'01)*, 2001.
4. Brian F. Chellas. *Modal Logic: An Introduction*. Cambridge University Press, Cambridge, 1980.
5. Robert Demolombe. To Trust Information Sources: a proposal for a modal logical framework. In Cristiano Castelfranchi and Yao-Hua Tan, editors, *Trust and Deception in Virtual Societies*. Kluwer Academic Publishers, Dordrecht, 2001.
6. Ronald Fagin, Joseph Y. Halpern, Yoram Moses, and Moshe Y. Vardi. *Reasoning about knowledge*. MIT Press, Cambridge, MA, 1995.
7. Sebastien Konieczny and Ramon Pino Perez. On the logic of merging. In *Proc. 6th Int. Conf. on Principles of Knowledge Representation and Reasoning (KR'98)*, pages 488–498. Morgan Kaufmann, 1998.
8. Sebastien Konieczny and Ramon Pino Perez. Merging with integrity constraints. In *Proc. 5th Eur. Conf. on Symbolic and Quantitative Approaches to Reasoning with Uncertainty (ECSQARU'99)*, number 1638 in LNCS. Springer, 1999.
9. Churn-Jung Liau. Logical Systems for Reasoning about Multi-agent Belief, Information Acquisition and Trust. In *Proc. ECAI 2000*, 2000.
10. M. Wooldridge and A. Lomuscio. A Computationally Grounded Logic of Visibility, Perception, and Knowledge. In *Logic Journal of the IGPL*, 9(2):273-288, 2001.

Agent-Oriented Language Engineering for Robust NLP

Vincenzo Pallotta and Afzal Ballim

Swiss Federal Institute of Technology - Lausanne
MEDIA group - Theoretical Computer Science Laboratory
IN F Ecublens, 1015 Lausanne, Switzerland
{Vincenzo.Pallotta, Afzal.Ballim}@epfl.ch

Abstract. The main goal of our work is to propose an agent-oriented framework for developing robust NLP applications. This framework provides means to compose analysis modules in a co-operative style. The idea is to encapsulate existing analysis tools and resources within software agents coordinated at a higher level using meta-knowledge. Agents can be activated concurrently and they should provide their linguistic competence depending on the application needs. The activation policy is determined by the context, by the domain knowledge and by performance constraints. At this level, co-ordination is computational logic-based in order to exploit known inference mechanisms for the decision support. This framework should be general enough to cope with other kinds of information sources, such as multimedia documents and with multimodal dialogue systems.

1 Introduction

The human ability in language interpretation is the result of thousands of years of evolution and cultural development. However also humans may have strong limitations in interpreting language in situations where they lack of contextual and domain knowledge or when they do not have the right competence of the language used. It is apparent that it is always possible to provide an interpretation of whatever kind of data. Not always one is able to provide the intended or the best approximating interpretation among the possible ones. This happens for humans and, why not, for artificial system. When switching to artificial systems, what is worth considering is that the human ability to provide approximate interpretations ranges from full understanding to complete ignorance. In addition, humans may overcome their limitation by their learning capabilities.

Processing unrestricted natural language is considered an AI-hard task. However various analysis techniques have been proposed in order to address specific aspects of natural language analysis. In particular, recent interest has focused on providing approximate analysis techniques (known as robust techniques), assuming that perfect analysis is not possible, but that partial results are still very useful. Lack of knowledge, uncertainty, vagueness, ambiguity, and misconception will have to be represented and considered in the interpretation process in order to handle linguistic applications "robustly".

A. Omicini, P. Petta, and R. Tolksdorf (Eds.): ESAW 2001, LNAI 2203, pp. 86-104, 2001.

Since interpretation of language is a subjective cognitive process, its formalization could be either a straightforward or a hard task depending on the taken perspective. Instead of committing us to a sort of closed world assumption where a failure is interpreted as the impossibility rather than a temporal inability due to the actual lack of specific knowledge, we consider lack of knowledge as *meta-knowledge* and its awareness may help us in judging the quality of the provided partial solution.

There are many ways in which the topic of *robustness* may be tackled: as a competency problem, as a problem of achieving interesting partial results, as a shallow analysis method, etc... They all show that no simple combination of "complete" analysis modules for different linguistic levels in a chain can give a robust system, because they cannot adequately account for real-world data. Rather, robustness must be considered as a system-wide concern where there must be a negotiation about which component can best solve a specific problem.

1.1 Modularity in Linguistics

We found inspiration for the development of our vision from recent trends in theoretical linguistics where modularity is being taken into greater consideration for language and discourse analysis of written text. In [24] some principles and methodological issues seem to provide us an underlying theoretical framework: instead of postulating a rigid decomposition of language into linguistic components (e.g. syntax, semantics, pragmatics), they propose to apply modularity to analysis by defining autonomous analysis modules, which in turn may consider different aspects of the language. The decomposition in linguistic levels may be still useful but it should not be considered as a pre-theoretical hypothesis on the nature of the language. Instead, analysis modules may take advantage of this decomposition to identify what features of the language should be considered. Modular independence (i.e. autonomy) is also a key issue. In order to provide a global modelling of the language, autonomous analysis modules can be composed by means of global meta-rules. To achieve this goal, modules should have a common interchange language and provide a global accessibility. The main advantages to methodological modularity in linguistics can be summarized in the following points:

- Extraction of language features at different granularities (i.e. phonemes, words, phrases, discourse, etc.).
- Co-existence of different analysis strategies that embody different principles.
- Coverage obtained as the composition of different (and possibly overlapping) language models and analysis techniques.

While the main goals in linguistics are the description of the language and the explanation of language phenomena, computational linguistics is more interested on how the above these methodological improvements may help in the design of NLP applications. Our vision can be outlined as follows.

retical and the practical account to the language processing will lead to more robust methods in NLP.

1.2 Language Engineering and Distributed NLP

There is an always-increasing interest towards distributed cooperative approaches. On the two extremes of this approach we have unsupervised and supervised coordination of intervening autonomous modules. For instance, the TREVI toolkit [12] provides an environment for the rational design of object-oriented distributed NLP application where the cooperation between modules is statically decided and dynamically coordnated by a dataflow-based object manager. In contrast the Incremental Discourse Parser architecture [16] assumes no predefined coordination schema but rather the spontaneous cooperation of well defined autonomous linguistic experts. The latter belong to the class of *blackboard* systems. A blackboard is a centralized data structure managed by a coordinator providing a framework where different expert modules cooperate to solve a problem. Modules eagerly scan the blackboard to see if there is a message that has been posted to it. The coordinator typically controls the access to the blackboard and thus the computation is not truly distributed.

Another attempt of designing a distributed architecture for NLP, which lies in the middle between the two above extremes, is the TALISMAN system [33]. TALISMAN is an agency where each agent has a specific linguistic competence. Agents are able to directly exchange information using an interaction language. Linguistic agents are governed by a set of local rules (called laws), which determine the local behaviour. A set of global (coordination) laws regulates the communication. TALISMAN deals with ambiguities and provides a distributed algorithm for conflicts resolutions arising from uncertain information. One agent may ask some expert agents for disambiguating between multiple interpretations and they may negotiate among each other in case of conflict between two or more experts.

The most remarkable example of multi-agent architecture for NLP is that of the system CARAMEL [30]. This system combines two different level of processing running in parallel: a *subliminal* and a *conscious* one. These two processes are designed as multi-agent systems but with different principles. The subliminal process works in background and it is left unsupervised. A cascade of simple modules elaborates the input written on an extended version of blackboard called the *sketchboard*. Each module involved in the subliminal process transforms the input present in the sketchboard and replies with a feedback message to the previous module in the cascade. There is no "output" from the system since the entire process may or not lead to a kind of "stability" where the input is not modified anymore. The subliminal process is linked to the conscious process via a short term memory where the result of the subliminal processing appears as an unconscious *pop-up*. This event triggers further processing which is managed by some other *controlled processes*. Controlled processes encapsulate linguistic experts. The conscious processing has a rational nature and it is build on the top of *reflective* multi-agent system. Reflectivity is intended as the ability of a system to use a representation of itself to reason about its own beha viour. The proposed architecture is recursive and allows building compound meta-systems coordinated by *meta-agents* aimed at controlling the execution of a set of domain-specific agents and where each domain-specific agent can itself be structured processing which is managed by some other *controlled processes.* Controlled processes encapsulate linguistic experts. The conscious processing has a rational nature and it is build on the top of *reflective* multi-agent system. Reflectivity is intended as the ability of a system to use a representation of itself to reason about its own behaviour. The proposed architecture is recursive and allows building compound meta-systems coordinated by *meta-agents* aimed at controlling the execution of a set of domain-specific agents and where each domain-specific agent can itself be structured as a compound meta-system. A meta-agent encapsulates additional knowledge to coordinate several agents providing a variety of services. Conscious treatment of the input is required when the subliminal system reaches three possible configurations: feeling of understanding (i.e. stability), feeling of ambiguity (i.e. instability) and feeling of contradiction (i.e. conflict with the stable result of the sketchboard and previous rationally computed analyses).

The design of Robust Dialogue Systems [1, 3] is nowadays one of the most challenging issues in NLP. Multi-agent systems seem to represent the natural framework both for the conceptual design and for the actual implementation of a Dialogue Management System, due to the fact that distributed models of computing allow us to design complex systems by means of reusable software components. As remarked in [2] the role played by the software infrastructure is a non-trivial one. When switching to mixed-initiative dialogue system where a more natural form of interaction is required, it is crucial to rely on autonomous, loosely coupled interacting components. This kind of architecture provides the necessary computational background for developing portable and reusable dialogue systems with a clear separation between discourse modelling and task/domain reasoning. This will certainly lead to a richer and more natural interaction with the user.

The above approaches to the design of NLP systems are all motivated by the need of modelling the system behaviour by means of *content information* rather than exclusively by general principles. Moreover the coexistence of multiple theories may help the system in finding the most appropriate analysis strategy (e.g. heuristically choosing which modules to use and how to compose them). Within the same theory, possibly implemented by an efficient algorithm, there is a variety of *tuning parameters* that may radically alter their performance and outcomes. The ability to determine a-priori which parameters are responsible for different types of analyses is a key factor for achieving adaptive NLP systems[1].

Software Engineering (SE) and in particular Natural Language Engineering (NLE) needs both theoretical and practical issue to be considered when adopting a design methodology [17]. For instance, the Natural Language Engineering architecture GATE[2] [18], although very useful for designing modular NLP systems, it doesn't seem directly exploitable for distributed NLP[3], because of a design that is optimised for information extraction: rigid module coupling and document transformation-based communication. However, the new release of GATE there is a claim for providing

[1] In NLP system the choice of a parameter could be also the selection of a suitable linguistic resource (or its subparts).

[2] General Architecture for Text Engineering.

[3] The new release of GATE will introduce agent-based language processing capabilities and a new approach to linguistic resources distribution.

agent-based language processing capabilities and a new approach to linguistic resources distribution in order to allow:

- the encapsulation of a range of alternative control regimes,
- consolidate the algorithmic resource integration mechanisms that currently live in GATE and extend those mechanisms to support distributed tools,
- add integration and management mechanism for linguistic resources such as lexicons.

In our opinion, these new features are not sufficient in order to exploit the agent-orientation as a real design methodology. Nevertheless GATE may provide us for a useful infrastructure for linguistic resources distribution and access.

1.3 Benefits of Agent-Oriented Language Engineering

Following [34] we can summarize the motivations for a adopting a multi-agent architecture for robust intelligent language analysis:

- *Distributed information sources*: knowledge and linguistic resources may be scattered over various (physical) locations. Access to multiple resources maybe mediated and rendered in a uniform way.
- *Shareability*: applications need to access several services or resources in an asynchronous manner in support of a variety of tasks. It would be wasteful to replicate problem-solving capabilities for each application. Instead it is desirable that the architecture supports shared access to agent capabilities and retrieved information.
- *Complexity hiding*: A distributed architecture allows specifying different independent problem-solving layers in which coordination details are hidden to more abstract layers.
- *Modularity and Reusability*: A key issue in the development of robust analysis application is related to the enhancement and integration of existing stand-alone applications. Agent may encapsulate pre-existing linguistic applications, which may serve as components for the design of more complex systems. Inter-agent communication languages improve interoperability among heterogeneous services providers.
- *Flexibility*: Software agents can interact in new configurations "on-demand", depending on the information flow or on the changing requirements of a particular decision making task.
- *Robustness*: When information and control is distributed, the system is able to degrade gracefully even when some of the agents are not able to provide their services. This feature is of particular interest and has significant practical implications in natural language processing because of the inherent unpredictability of language phenomena.
- *Quality of Information*: The existence of similar analysis modules able to provide multiple interpretation of the same input offers both 1) the possibility of ensuring the correctness of data through cross-validation and 2) a mean of negotiating the best interpretation(s).

The importance of adopting an agent-oriented methodology from a linguist point of view lies on the fact that the design of an NLP system able to deal with real language phenomena can be made at knowledge level rather than implementation level. The properties of the system can be declaratively specified as sets of global meta-rules integrating the specialised linguistic processors. Following Petric [27], Agent-Based Software Engineering will:

- Reduce the need of detailed specification
- Handle error gracefully

- More abstraction in system building
- Semantic defined as *emergent behaviour*.

These nice features are exactly those a language engineer is looking for.

In conclusion, we can summarize our methodology proposal for distributed NLP with the following schema:

- Planning (with capability brokering)
  - o Generate a sequence of task to be performed by expert linguistic agents
  - o Implicit control of the execution
  - o Recover from partial failures
  - o Plan recipes to be activated by contextual information
- Decentralized coordination
  - o Teamwork formation
  - o Multiple/different roles for linguistic agents depending on
  - o Types of analysis
  - o Stages of the analysis
  - o Performance constraints

2 The HERALD Project

In the HERALD[4] project we are aimed at implementing robust applications for natural language processing as the result of a design process that takes into account different computational perspectives and allows the combination of multiple problem-solving techniques. The Multi-Agent Systems (MAS) programming model seemed to naturally meet our requirements.

2.1 Agent Oriented Programming

The *Mentalistic Agent* model seemed to be the most appropriate since it allows us to design multi-agent systems where each agent may build and manage its own knowledge base and it is able to use it for taking decisions. We chose a rather simple model, the Agent Oriented Programming (AOP) [32], where the design of agents is made by

[4] HERALD stands for Hybrid Environment for Robust Analysis of Language Data.

AOP is the theoretical foundation of a commercial architecture AgentBuilder[5] for the design of software agents that we adopted as the development platform in this project. In AgentBuilder agent can be specified by set of *commitment rules* specifying where actions are executed matching both conditions on the received messages and the mental state. Messages are encoded using the KQML agent communication language [22], a high-level language intended to support interoperability among intelligent agents in distributed applications. Message conditions are evaluated examining the content of KQML messages. AgentBuilderis an *open system*, in which additional processing components, called *Project Accessory Classes* (PAC), can be easily added. They are classes used by the agents as a mechanism to interact with its environment. PACs can be coded in Java or other programming languages supporting the Java Native Interface (JNI). PACs provide an ideal mechanism for wrapping legacy code and using a PAC the agent designer creates *Ontologies* that are collections of object classes an agent can use to construct its mental state.

2.2 Extensions of AOP in HERALD

In this section we review some of the additional features we added to AOP to support the design of NLP applications.

- *Mental State management in ViewGen.* The ViewGen system [10, 11] is intended for modelling agents' mutual beliefs. An environment represents each agent and it may use nested environments to represent other's agent propositional attitudes. Relationships between environments can be specified hierarchically or by the explicit mapping of entities. Each environment has associated an axiomatization and a reasoning system. This system has been integrated with AOP in order to provide a more sophisticated method for the agents' mental states management.
- *Capability Brokering.* A software layer for the processing of the KQML primitives for agent recruitment in multi-agent systems has been implemented. The *broker* is the agent who has the knowledge about the capabilities of different "problem-solving" agents who "advertise" their capabilities by means of a *Capability Description Language* (CDL) [36].
- *The IRC Facilitator.* We implemented the IRC[6] PAC that allows an agent to post messages on a shared message-board. The IRC PAC allows the construction of agents capable of connecting to an IRC server that offers routing and naming facilities. IRC can be viewed as an alternative agent communication infrastructure based on a hub topology that provides at the same time peer-to-peer and broadcasted message passing. The IRC facilitator may serve also as a platform to implement blackboard-based applications within AOP.

---

[5] `http://www.agentbuilder.com`.

[6] IRC stands for International Relay Chat. IRC is a widely used Internet chat server.

2.3 HERALD Agent Logic

Recent trends in designing and developing intelligent systems have shown that there is an increasing need of incorporating elements of *rationality* in the programming languages. In particular, the design of multi-agent systems requires an account of the quality of choice making. The motivations for representing the knowledge about the system's features are:

- the heterogeneity of the knowledge sources,
- the diversity of problem solving capabilities supported by involved actors,
- the presence of multiple roles that may be covered by the same service providers.

Formal reasoning about agents' behaviour and social interaction allows us to integrate, coordinate, and monitor planning and plan execution, and to incrementally improve the efficiency and robustness of the multi-agent analysis system.

In its original formulation AOP has been defined in a quite informal way, without making explicit the relationship between its actual implementation as a programming language (AGENT0 [31]) and its underlying modal logic. Moreover there is a fundamental mismatch between the AOP modal logic and AGENT0: whereas in the logic commitments are represented as obligations to a fact holding, in the programming language an agent commits to actions to change its mental state. The formalisms we consider for the design of a new logical framework for AOP are the following:

- *Temporal Belief Logic.* The Wooldridge's Temporal Belief Logic (TBL) [37] aims to model the agent's mental state evolution. The main reason for doing so is that one can exploit this information in order to perform two different things:
  1. Formally verify the multi-agent system properties (e.g. safety, liveness, fairness, etc.).
  2. Provide a meta-language to be used at co-ordination level for reflecting the multi-agent system behaviour.
- *Capability Description Language.* The utility of a capability brokering mechanism is apparent if our goal is to conceive an open architecture where modules can be plugged-in without any direct intervention. We need a formal machinery to specify the processing power of the modules encapsulated by the agents.
- *Action Logic.* We contributed to the design of the Fluent Logic Programming (FLP) action language and to the implementation of its proof procedure [25, 26]. The nice feature of FLP is that it allows to model incomplete information about world states, which can be inferred on demand in a particularly efficient way. Although not extremely expressive, it can suitably represent AOP agents' interaction. Moreover its capacity of dealing with incomplete knowledge is definitely useful to model agents' hypothetical behaviours.
- *Viewpoints logic.* ViewFinder [4] is a framework for manipulating *environments.* Environments (or views, or partitions, or contexts) are aimed to provide an explicit demarcation of information boundaries, providing methodological benefits (allowing one to think about different knowledge spaces), as well as processing them (allowing for local, limited reasoning, helping to reduce combinatorial problems,

The nice feature of FLP is that it allows to model incomplete information about world states, which can be inferred on demand in a particularly efficient way. Although not extremely expressive, it can suitably represent AOP agents' interaction. Moreover its capacity of dealing with incomplete knowledge is definitely useful to model agents' hypothetical behaviours.

- *Viewpoints logic.* ViewFinder [4] is a framework for manipulating *environments.* Environments (or views, or partitions, or contexts) are aimed to provide an explicit demarcation of information boundaries, providing methodological benefits (allowing one to think about different knowledge spaces), as well as processing them (allowing for local, limited reasoning, helping to reduce combinatorial problems, etc.). The ViewFinder framework provides the foundations for the following issues related to the manipulation of environments:
  - Correspondence of concepts across environments
  - Operations performed on environments
  - Maintenance of environments.

  ViewFinder is the theoretical foundation for ViewGen and in its full generality and it can be used for modelling multi-dimensional reasoning systems.

The integration of the above formal theories into a unique logical framework allows us to capture the abstract properties of AOP. The AOP adopted implementation (e.g. AgentBuilder) provides an excellent framework for software composition but it was not thought as a logical framework for representing abstract properties of the designed multi-agent systems. We propose a common logical language to express:

- agents' mental state evolution by means of a Temporal Belief Logic,
- agents' mutual beliefs by means of a logic of viewpoints (ViewGen),
- agents' interaction with the environment by means of an action Logic (FLP),
- agents' capabilities by means of a Capability Description Language.

ViewFinder provides the supporting backbone for the integration of different knowledge representation languages. It is worth to observe that we concentrate on a special case of AOP and this simplifies in part the work. This may also represent a first step in the implementation of ViewFinder as fully-fledged knowledge representation framework.

In order to exploit information represented using the HERALD Agent Logic at the agents' decision support level (e.g. coordination), we need to provide a suitable proof theory. Actually ViewFinder is characterized by its operational semantics expressing only how information can be *ascribed* and *percolated* between environments and how shared entities can be referenced in different environments using different names. The issue on how to support the cross-environmental reasoning is left open. We adopted the ViewFinder underlying operational semantics and we mapped it into a multi-context logic like one of those proposed by [13], which serves also as a meta-language for specifying the axiomatization of the Agent Logic meta-theory. A meta-interpreter for the *HERALD Agent Logic* is currently under study and it will be implemented as a CHR constraint solver [20].

3 Design of Robust NLP Applications in HERALD

The HERALD framework aims to fulfil both *software engineering* and *computational linguistics* requirements providing a higher-level, formal and configurable design environment. In order to be able to deal with robustness at various linguistic levels we realize that this is only possible if we could rely on a formal framework that allows us to conceive and experiment ways of combining linguistic modules.

3.1 Past Experiences

Robust analysis of text was the main concern of the previously research project ROTA[7]. The main achievement of that project was the conception of a robust parser compiler (e.g. LHIP[8] [9]) and its integration with heterogeneous knowledge sources. Indeed the composition of different sources of linguistic competence leads to a more adaptive behaviour of the overall system. However, the composition strategy must be flexible enough to avoid both the introduction of erroneous early commitments and delayed decisions. In the former case the error propagates affecting the final outcome. In the latter it could dramatically decrease the performance.

A summary of the improvements in the LHIP is summarized in [23]. The underlying idea of the LHIP system is very attractive since it allows one to perform parsing at different arbitrary levels of "shallowness". On the other hand, implementation is not completely satisfactory. Granted that LHIP had reached a limit in computational feasibility, it can be envisioned that its role in a NLP system would be:

- Chunk extraction
- Implementation of semantic grammars
- Concept spotting
- Extraction of dialogue acts

This project has shown the viability of the structural analysis techniques that we are developing, but it has also shown their shortcomings. Developing and integrating semantic analysis methods, which will be a crucial and fundamental advancement in this field, can best address these shortcomings.

We recently focused our attention on the integration of multiple domain and linguistic knowledge sources built on heterogeneous theories in order to obtain robust analysis. This kind of integration has already been proved to be fruitful in our previous project ISIS[9] [5] where rule-based and statistically driven methods where used to implement analysis modules at different linguistic levels. In this project [14] we proposed a framework for designing grammar-based procedure for the automatic extraction of the semantic content from spoken queries.

The availability of a large collection of annotated telephone calls for querying the Swiss phone-book database (i.e the Swiss French PolyPhone corpus [15]) allowed us

[7] ROTA stands for Robust Text Analysis.

[8] LHIP stands for Left-corner Head-driven Island Parser.

[9] ISIS stands for Interaction through Speech with Information Systems.

to test our findings from the project ROTA. Starting with a case study and following an approach, which combines the notions of fuzziness and robustness in sentence parsing, we showed we built practical domain-dependent rules, which can be applied whenever it is possible to superimpose a sentence-level semantic structure to a text without relying on a previous deep syntactical analysis. In the case of particular dialogue applications where there is no need to build a complex semantic structure (e.g. word spotting or extracting) the presented methodology may represent an efficient alternative solution to a sequential composition of deep linguistic analysis modules. Another critical issue is related to overall robustness of the system. In our case study we tried to evaluate how it is possible to deal with an unreliable and noisy input without asking the user for any repetition or clarification.

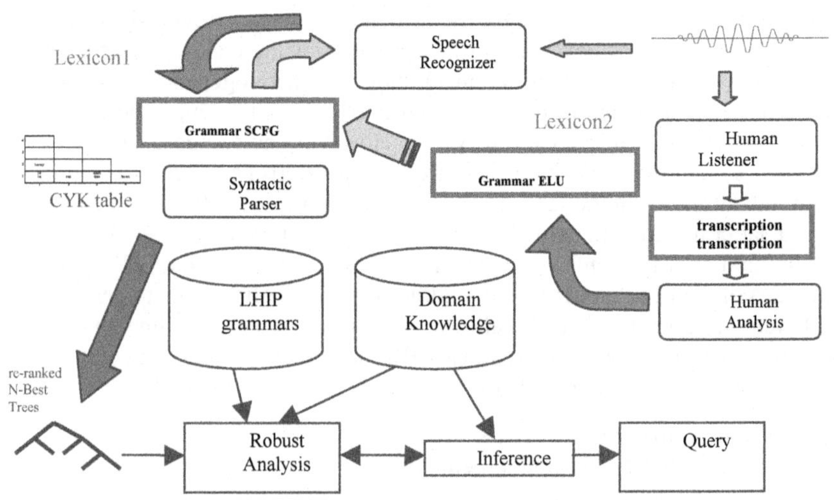

Fig. 1. The ISIS Architecture.

We realized that the architectural rigidity reduces the effectiveness of the integration since the system cannot decide dynamically *when* and *how* the cooperation should happen. Nonetheless our final opinion about the ISIS project is that there are some promising directions applying robust parsing techniques and integrating them with knowledge representation and reasoning.

Both in ROTA and ISIS we didn't concentrate on software engineering aspects of the overall analysis process. Parsing will now be considered as an important step in the analysis of natural language data, but not the only one. Its importance and centrality may change depending on the application types and domains. Moreover, its role among other processing modules may vary dynamically during the analysis. We decided to leave robust parsing and improvement of LHIP aside from our investigations until the complete set up of the HERALD architecture.

3.2 Robust Analysis Techniques

We identified the some main issues that we wanted to address in our scientific investigation about robustness in analysis of natural language data. We are going to outline how the HERALD project contributes to each of them.

Extending Coverage. When switching from a classical sequential linguistic architecture to a distributed one, the problem of coverage must be rephrased. In the case of parsing, coverage can be improved by enlarging the grammar or providing constraints relaxation mechanisms and keeping at same time the induced over-generation as little as possible. In a NLP distributed architecture where the parser is one of its components, coverage in the classical sense is always guaranteed. We may produce interpretations ranging from the empty to the intended one.
We consider a nice feature of the LHIP parser that allows the selective activation of subset of the grammar. The robust behaviour is represented by the ability of improving coverage without any direct modification of the grammar in use. By the simple modification of the coverage parameters we can decide whether to perform strict parsing, chuck parsing, shallow parsing or word spotting. In a distributed environment we can also imagine to run multiple instances of the LHIP parser with different coverage, providing different types of contribution to the overall interpretation process.

Improving Efficiency. A distributed architecture allows us to launch multiple instances of the same parser with different parameters and constraints (e.g. coverage thresholds, grammars, parsing strategies, look-ahead, timeouts etc.) and evaluate the obtained results.

Disambiguation. Incremental interpretation allows us to solve ambiguities when they show up. Ambiguities can be solved using accumulated contextual information at different linguistic levels and by *domain-driven semantic filtering* [8]. The latter technique allows us to rule out unintended interpretations by means of world knowledge and expectations using general or specific knowledge bases. In this first phase we investigated the possibility of encapsulating existing Ontologies and knowledge bases within agents in our architecture.

Approximation. In our published works [6, 7], we proposed the notion of weighted semantic parsing with LHIP. We made use of a "light-parser" for actually doing sentence-level semantic annotation The main idea comes from the observation that annotation does not always need to rely on the deep structure of the sentence (e.g. at morpho-syntactic level). In some specific domains it is possible to annotate a text without having a precise linguistic understanding of its content. A LHIP parse may easily produce several multiple analyses. The main goal of introducing weights into LHIP rules is to induce a partial order over the generated hypotheses and exploit it for further selection of k-best analysis. Each LHIP rule returns a weight together with a term, which will contribute to build the resulting parsing structure. The confidence factor for a pre-terminal rule is assigned statically on the basis of the domain knowledge, which allows us to find cue-words within the text.

3.3 Incremental Robust Interpretation

The need of having a robust treatment of users input is amplified by intrinsic recognition errors induced for example by an Automatic Speech Recognisers (ASR). A viable approach is to combine different levels of shallowness in the linguistic analysis of utterances and produce sets of ranked utterance's interpretations. Competing analyses maybe compared with respect to the confidence levels we assign to the producing linguistic modules (or the macro-modules that encapsulate some complex processing). For instance, multiple instances of the robust parser LHIP can be launched with different sub-grammars. They generate different analyses that can be taken into account for the extraction of the best interpretation. The agent that encapsulates the LHIP parser interacts with different linguistic modules in order to support contextual disambiguation and domain reasoning to reduce the search space. In general we consider the following robust analysis components to be encapsulated by AOP agents:

Partial and adaptive parsing. Linguistic theories are aimed to model an idealized language. Most of them propose a generative approach: the language modelled can be generated by a set of production rules. When dealing with spoken and written language we may face with phenomena that cannot be captured by general principle. Instead of enlarging the set of generative rules to cope with extra-grammatical phenomena our robust parser LHIP allows us to approximate the interpretation adopting some heuristic in order to gather sparse possibly correct analysis of sub-constituents. Heuristics can be used also to automatically determine the domain-related features of the language to be considered, and thus select the appropriate sub-model of the language.

Underspecified semantics. Studies have shown that in some cases it is necessary to delay the decision of how to build a semantic interpretation up to the moment when additional contextual information is obtained. Once we are able to construct a logical form for a given utterance (possibly exploiting already accumulated additional information), it could be the case that spurious ambiguity remains which can be solved only at later stages of the analysis. We adopted some methods for representing semantic ambiguity (e.g. multiple quantifiers' binding) in an efficient and compact way [28].

Abductive discourse interpretation. Interpretation problems like reference resolution, interpretation of compound nominals, the resolution of syntactic ambiguity and metonymy and schema recognition require a adaptive inference if done at the level of semantic logical form [21]. The main problem with system that allows open interpretations is that of the combinatorial explosion: multiple interpretations will induce an exponential growing of interpretations at later stages of processing. The worst case is when we sequentially combine modules producing multiple interpretations without any selection mechanism that reduces the search space and keeps the computation tractable. Unfortunately, ambiguities arisen at a certain linguistic level could be solved only with the information provided at higher levels. If we do not consider an incremental approach we are not able to exploit this information: a transformational approach would force us to generate all the interpretation of one level and feed the subsequent linguistic module with all the produced interpretations. In the best case we are able to rank the hypotheses and select the n-bests. Once the utterance has been interpreted (both multiple or underspecified) the Interpretation Manager is able to generate the problem-solving act as KQML message. Only plausible recognized communicative acts should be "assimilated" in Discourse Context meaning that contextual interpretation and pragmatic reasoning is required in order to resolve ellipses and anaphora.

3.4 Agent-Oriented NLP Architecture

In a modular NLP system the choice of what information the linguistic modules may exchange is a critical issue. In particular the problem become more complex when a module does not know in advance what kind of processing is required on the received data. For instance a module may require specific processing type constraints or tell what resources must be used. The richness of the possible accessory information that could be sent together with the object data imposes a better way of encoding. KQML allows us to wrap data into messages in a standardized way. It also allows us to specify the language required to interpret the message content. The advantage of using an agent communication language like KQML is both at specification and implementation level. First we can define abstract cooperation protocols between modules that are mapped later into specific processing threads. Second, simply changing its communication interface one can reuse the same NLP processor to accomplish different tasks. For instance, we can encapsulate our flexible robust parser LHIP [9] into an agent and ask for different level of coverage or asking as result the chart of partial parses.

It is apparent that we need to model the quality of service offered by a linguistic expert agent. We consider including in our capability descriptions the following information:

- Reliability of the analysis
    - Confidence factors
- Performance
    - Evaluated precision and recall
    - Efficiency
- Cross-domain transferability
    - Reuse of trained linguistic modules
    - Ontologies
    - Multilingualism
- Availability of the linguistic resource/processors
    - Usage restrictions
    - Concurrent access

We would also like to reproduce in our architecture the dataflow-oriented threading mechanism in the TREVI architecture by possibly mapping it to a suitable usage of *typed messages*. Here the coordinating agent that holds a dynamically constructed view of the system takes the role of the Flow-Manager and the flow-graph is represented by a nested environment representation in ViewFinder. Several agents having a general internal structure compose the HERALD architecture. An *HERALD reasoning agent* encapsulates an inference engine and a local knowledge base. A core set of rules is devoted to the management of its mental state and to the interface with the coordination agent (i.e. the *Knowledge Mediator Agent*). Special purpose agents are included for the *multi-modal input management* and for specific processing of the input (e.g. *Interpretation Manager*). In Figure 1 we describe the HERALD architecture.

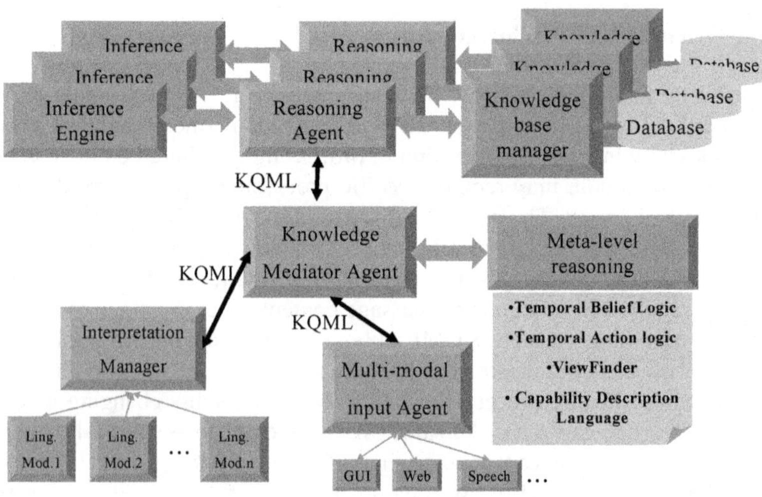

Fig. 2. The HERALD General Architecture for NLP Applications.

The Knowledge Mediator (KM) agent is responsible of the coordination of modules. By means of meta-rule expressed as AOP commitment rules, it can take decisions on how to assign tasks to the NLP expert agents. The KM agent embodies a broker, which looks for available resources (both linguistics and computational) and plans the future course of actions depending on the capabilities of the recruited agents. The idea is that the choice of a linguistic or a computational resource may depend on the current context. Since the final goal of a NLP is to provide an interpretation of the input, the way this task is performed may vary upon the previous collected information. An NLP system should adapt its computational effort to different type of situations where already present information may create stronger assumption on the input to be processed.

3.5 Text Understanding in HERALD

We are considering now a more general NLP application that we chose for testing the HERALD design environment; *text understanding* requires the integration between linguistic information and knowledge of the world. The integration may be required at any level of linguistic description. GETARUNS (General Text And Reference Understanding System) [19] is an existing system for text analysis and understanding which embodies this principle. Although its generality, it is oriented towards the treatment of narrative texts. GETARUNS is able to build a model world where relations and entities introduced and referred to in the text are asserted, searched for and ranked according to their relevance. The system is composed of the following linguistic modules:

- Top-down LFG syntactic parser
- Sentence-level anaphoric binding module
- Discourse-level anaphora resolution module
- Semantic modules for logical form generation and quantifiers raising.
- Sentence-level temporal and aspect analysis module
- Discourse-level temporal analysis module
- Conceptual representation and reasoning module
- World model builder module

GETARUNS is conceptually a modular architecture but it is implemented as an almost monolithic sequential PROLOG program. In order to verify the effectiveness of our proposed architecture/methodology we will approach the *agentification* of GETARUNS as a reverse engineering problem. Undergoing experimentations will deal with the design of novel analysis strategies based on the possibility offered by the HERALD architecture of having a rule-based control of the dynamics of computations. We will implement heuristic as to dynamically change the dataflow and the linguistic processors' parameters.

4 Conclusions

The challenge undertaken in the HERALD project is in the ability to gather and "intelligently" process relevant information coming from different knowledge sources. This can be guaranteed by improving the flexibility of these sources processing by introducing a higher level of rationality in our software artefacts. We argue that this is the key to robustness of future NLP applications. We can summarize the HERALD methodological features as follows:

- Rule based specification of system's modules composition. Modules should be loosely coupled and allow dynamic reconfiguration of the system topology. Composition rules should account of types of data object that modules are supposed to exchange.

- Dynamic task assignment based on contextual information. Coordination modules should be able to access information about others modules capabilities, evaluate their performance and select among best response when multiple modules are activated in parallel to accomplish the same goal.

- Logic based decision support. The coordination decisions should be taken rationally. This means that in our architecture we should be able to design agents ranging from those having reactive behaviour to deliberative ones (including BDI and reflective agents).

As in TREVI, DPA, TALISMAN, CARAMEL and other systems, we propose to move towards a distributed solution in which software agents encapsulate flexible analysis modules. One the one hand we need to specify at an abstract level how modules interact and the dataflow. On the other hand we need to map the abstract specification to a concrete architecture. In contrast to the TREVI approach we do not require an initial commitment on the architecture topology that may also vary during the processing. The similarity between the HERALD architecture and the CARAMEL system are apparent. In the HERALD architecture reflexivity is obtained by the representation of viewpoints in the controller agent's mental space. In contrast our architecture does not directly address the issue of unconscious-conscious dichotomy, even if it provides the support for implementing shared memories spaces (e.g. blackboards). The choice of the AOP formal model of programming will allow us to declaratively design the behavioural features of our systems and to use formal methods to prove their properties.

The current setting of the HERALD project allows us to identify two complementary perspectives of the same problem that have been addressed in our investigations, namely: *Robust Analysis Techniques* and *Robust Language Engineering* . The methodological improvements introduced in HERALD may constitute an important step towards the design of robust analysis systems capable of dealing with highly unstructured but rich of semantic content documents. The assumption on robust parsing was that partial results could be often much more useful than no result at all, and that an approximation to complete coverage of a document collection is more useful when it comes with indication of how to complete it. We now make a further step and consider the analysis process as composed of different overlapping sub-tasks where robust natural language processors are the building blocks of what we consider the future's NLP applications. Only if we are able to tell our programs how to combine them to extract the intended meaning we will succeed in design real robust NLP applications.

References

1. Allen, J., Byron, D., Dzikovska, M., Ferguson, G., Galescu, L. and Stent, A. Towards a Generic Dialogue Shell. *Natural Language Engineering,6(3)* . 1-16.
2. Allen, J., Ferguson, G. and Stent, A., An architecture for more realistic conversational systems. in *Proceedings of Intelligent User Interfaces 2001 (IUI-01)*, Santa Fe, NM, 2001.
3. Allen, J.F., Miller, B., Ringger, E. and Sikorski, T. A Robust System for Natural Spoken Dialogue. in *Proc. 34th Meeting of the Assoc. for Computational Linguistics* , Association of Computational Linguistics, June, 1996.
4. Ballim, A. ViewFinder: A Framework for Representing, Ascribing and Maintaining Nested Beliefs of Interacting Agents *Computer Science Department* , University of Geneva, Geneve, 1992.
5. Ballim, A., Chappelier, J.-C., Pallotta, V. and Rajman, M., ISIS: Interaction through Speech with Information Systems. *Proceedings of the 3rd International Workshop, TSD 2000*, (Brno, Czech Republic, 2000), Springer Verlag, 339-344.
6. Ballim, A. and Pallotta, V., Robust parsing techniques for semantic analysis of natural language queries. *Proceedings of VEXTAL'99 conference* (Venice, I, 1999).

10.Ballim, A. and Wilks, Y. *Artificial Believers*. Lawrence Erlbaum Associates, Hillsdale, New Jersey, 1991.

11.Ballim, A. and Wilks, Y. Beliefs, Stereotypes and Dynamic Agent Modelling. *User Modelling and User-Adapted Interaction, 1* (1), 1991, 33-65.

12.Basili, R., Mazzucchelli, M. and Pazienza, M.T., An Adaptive and Distributed Framework for Advanced IR. in *6th RIAO Conference (RIAO 2000)*, Paris, 2000.

13.Bonzon, P., A Refective Proof System for Reasonning in Contexts. in *AAAI 97 National Conference*, Providence, Rhodes Island, 1997.

14.Chappelier, J.-C., Rajman, M., Bouillon, P., Armstrong, S., Pallotta, V. and Ballim, A. ISIS Project: final report, Computer Science Department - Swiss Federal Institute of Technology, 1999.

15.Chollet, G., Chochard, J.-L., Constantinescu, A., Jaboulet, C. and Langlais, P. Swiss French PolyPhone and PolyVar: Telephone speech database to model inter- and intra-speaker variability, IDIAP, Martigny, April, 1996.

16.Cristea, D., An Incremental Discourse Parser Architecture. in *Second International Conference - Natural Language Processing - NLP 2000*, Patras, Greece, 2000, Springer.

17.Cunningham, H., Bontcheva, K., Tablan, V. and Wilks, Y., Software Infrastructure for Language Resources: a Taxonomy of Previous Work and a Requirements Analysis. in *Proceedings of LREC 2000*, Athens, Greece, 2000.

18.Cunningham, H., Gaizauskas, R. and Wilks, Y. *A General Architecture for Text Engineering (GATE) -- a new Approach to Language Engineering R & D.* University of Sheffield, 1995.

19.Delmonte, R., Parsing with GETARUNS. in *Proceedings of the "7eme conférence annuelle sur le Traitement Automatique des Langues Naturelles TALN 2000*, Lausanne, 2000.

20.Fruehwirth, T. Theory and Practice of Constraint Handling Rules. *Journal of Logic Programming, Special Issue on Constraint Logic Programming (P. Stuckey and K. Marriot Eds.)* (Vol 37(1-3)). 95-138.

21.Hobbs, J., Stickel, M., Appelt, D. and Martin, P. Interpretation as Abduction. *Artificial Intelligence, 63(1-2)* (1-2). 69-142.

22.Labrou, Y. and Finin, T. A Proposal for a new KQML Specification, Computer Science and Electrical Engineering Department, University of Maryland Baltimore County, Baltimore, MD 21250, 1997.

23.Lieske, C. and Ballim, A. Rethinking Natural Language Processing with Prolog. in *Proceedings of Practical Applications of Prolog and Practical Applications of Constraint Technology (PAPPACTS'98)*, Practical Application Company, London,UK, 1998.

24.Nolke, H. and Adam, J.-M. (eds.). *Approches modulaires: de la langue au discourse.* Delachaux et niestlé, Lausanne, 1999.

25.Pallotta, V., A meta-logical semantics for Features and Fluents based on compositional operators over normal logic-programs. in *First International Conference on Computational Logic*, London, UK, 2000, Springer.

26.Pallotta, V., Reasoning about Fluents in Logic Programming. in *Proceedings of the 8th International Workshop on Functional and Logic Programming*, Grenoble, FR, 1999, University of Grenoble.

27.Petrie, C.J. Agent-Based Software Engineering. in Ciancarini, P. and Wooldridge, M.J. eds. *Agent-Oriented Software Engineering*, Springer Verlad, Limerick, 2000.

28.Poesio, M., Disambiguation as (Defeasible) Reasoning about Underspecified Representations. in *Papers from the Tenth Amsterdam Colloquium*, Amsterdam, 1995.

29.Rao, A.S. and Georgeff, M.P. Modeling Rational Agents within a BDI-Architecture. in Allen, J., Fikes, R. and Sandewall, E. eds. *Proceedings of the 2nd International Conference on Principles of Knowledge Representation and Reasoning*, Morgan-Kaufmann, 1991, 473-484.

30.Sabah, G. Consciousness: a requirement for understanding natural language. in S. O. Nuallàin, P.M.K., E.M. Aogàin ed. *Advances in Consciousness research*, John Benjamins, Amsterdam, 1997, 361-392.

31.Shoham, Y. AGENTO: A Simple Agent Language and Its Interpreter. in *Proceedings of the Ninth National Conference on Artificial Intelligence (AAAI'91)*, AAAI Press/MIT Press, Anaheim, California, USA, 1991, 704-709.
32.Shoham, Y. Agent-Oriented Programming. *Artificial Intelligence, 60* (1). 51-92.
33.Stefanini, M.-H. and Demazeau, Y., TALISMAN: a multi-agent system for Natural Language Processing. in *IEEE Conference on Advances in Artificial Intelligence, 12th Brazilian Symposium on AI*, Campinas, Brasil, 1995, Springer, 310-320.
34.Sycara, K. and Zeng, D. Coordination of Multiple Intelligent Software Agents. *International Journal of Cooperative Information Systems, 5(2-3)*.
35.Thomas, S.R. The PLACA Agent Programming Language. in Wooldridge, M.J. and Jennings, N.R. eds. *Proceedings of the ECAI-94 Workshop on Agent Theories, architectures and languages: Intelligent Agents I*, SV, Berlin, 1995, 355-370.
36.Wickler, G.J. Using Expressive and Flexible Action Representations to Reason about Capabilities for Intelligent Agent Cooperation *Computer Science Department*, University of Edimburgh, Edimburgh, 1999.
37.Wooldridge, M.J. On the Logical Modelling of Computational Multi-Agent Systems *Department of Computation*, UMIST, Manchester, UK, 1992.

Extending a Logic Based One-to-One Negotiation Framework to One-to-Many Negotiation

Paolo Torroni[1] and Francesca Toni[2]

[1] DEIS, Università di Bologna
Viale Risorgimento 2, 40136 Bologna, Italy
ptorroni@deis.unibo.it
[2] Department of Computing, Imperial College
180 Queens Gate, SW7 London, UK
ft@doc.ic.ac.uk

Abstract. [13] presents a logic-based approach to multi-agent negotiation. The advantages of such approach stem from the declarativeness of the model, which allows to formulate and prove some interesting properties (such as termination and convergence of a protocol), to the possibility of identifying and combining varieties of agents, implementing different negotiation policies, and of forecasting the behavior of a system with no need for simulation. The work introduces a language for negotiation that allows to cater for two agent dialogues, in a one-to-one negotiation setting. Auctions are an example of one-to-many negotiation mechanisms, where agents try to maximize their profit by buying items in competition with other parties, or selling them to crowds of bidders. In this paper, we show how the negotiation framework of [13] can be extended to accommodate a suitable negotiation language and coordination mechanism (in the form of a shared blackboard) to tackle one-to-many negotiation.

1 Introduction

Autonomous Agents and Multi-Agent Systems (MAS) have represented a hot topic of Computer Science disciplines for the past ten and more years. They have often been adopted as a metaphor to model autonomous entities capable of interacting and being part of organizations or societies. The idea of making them intelligent asked for the contribution of disciplines such as Artificial Intelligence (AI) [16]. Logic-based AI has been playing a foreground role in the MAS community since the very beginning, e.g. with the advent of the BDI model for agent beliefs, desires, and intentions [9].

Recently, computational logic has started to make a relevant contribution to the development of MAS [10,14]. Indeed, computational logic-based formalisms are a powerful way to model and implement the agents' knowledge and the reasoning of the agent, that uses and possibly updates such knowledge. All these ingredients suit well a MAS context, which is normally open, dynamic, and unpredictably evolving. Computational logic can contribute to modeling both individual agents and societies of agents, and to providing operational semantics that can be straightforwardly used as a basis to implement a system.

In [13,12], Sadri, Toni and Torroni introduced a computational logic-based approach to agent dialogue for negotiation of resources. Negotiation is one of the main research

A. Omicini, P. Petta, and R. Tolksdorf (Eds.): ESAW 2001, LNAI 2203, pp. 105–118, 2001.

streams in MAS, due both to the wide range of possible applications, spanning electronic markets, task reallocation, and distributed resource management, just to cite some, and to its strong commercial interest. In a logic setting, agents are given a logic-based knowledge representation, including goals, intentions, and beliefs. When an agent is missing a resource, it negotiates with another agent, in a dialogue-based framework, requesting the missing resource and carrying out the dialogue by means of utterances, or *dialogue primitives*. In the course of a dialogue, both agents could possibly modify their own intentions, for instance as a consequence of a 'better' (i.e. less expensive) plan that could arise during the dialogue. Agents decide which primitive to utter based upon a computational logic-based proof procedure, that treats dialogue primitives as 'hypotheses' (or 'abducibles'). Such hypotheses are singled out so that some given 'negotiation policies' are enforced. The policies are represented as 'integrity constraints' that need to be satisfied, as in databases and abductive reasoning.

In realistic applications, negotiation mechanisms need to cater for and exploit the plurality of agents in societies. The agents looking for resources might want to choose among several marketplaces, the agents providing resources might want to choose among different possible buyers, and all agents, in general, might want to play as buyers and sellers, in different contexts, according to their subjective needs and to the objective situation. In this respect, *auctions* can be seen as negotiation patterns where agents do not deal with each other in pairs, but rather compete in crowds for the achievement of their individual goals. In auctions, agents try to maximize their profit by buying items in competition with other parties, or selling them to selected agents in crowds of bidders.

In this paper, we show how to tackle one-to-many negotiation within the one-to-one negotiation framework of [13]. The contribution of the paper is in the introduction of a logic approach to automated auctions, and in particular in the extension of an existing framework and in the proposed implementation of the English auction protocol by means of a program written in the extended language. One of the most innovative aspects of such approach is that it is operational, i.e., the logic that describes the agent knowledge is a program that is implementable to build agent applications, whose properties can be studied and proven. We extend the negotiation language, and we introduce a suitable coordination mechanism (via a shared blackboard). Although this is ongoing work, we believe that such approach is a promising starting point for building up an integrated logic-based negotiation framework, within which we aim to define and prove properties that rule the marketplace, without resorting to simulation.

The paper is organized as follows: in Sections 2 and 3 we briefly discuss the one-to-one negotiation language and framework, respectively, introduced in [13]. In Section 4 we introduce the concept of auctions, referring to some classical auction protocols in the literature. Starting from the individual protocols, we draw a common schema driving our choices in the design of the extended language. In Section 5 and 6 we show how to extend the framework to tackle auctions, and we illustrate the extension by realizing a particular protocol, the English auction. In Section 7 we draw some conclusions.

2 A Language for Negotiation

In this section we will briefly sketch the language for negotiation adopted in [13] and give an example of a two-agent negotiation dialogue. In the next section, we will give a flavor of how such dialogue can be generated within a computational logic setting.

In the following dialogue, inspired by [8], agent a asks agent b for a resource (a $nail$), needed to carry out a task (to hang a picture). Being refused the requested resource, a asks b the reason why, aiming at acquiring additional information that could help a to bargain for the resource or find an alternative resource to carry out the task.

$tell(a, b, request(give(nail)), 1)$
$tell(b, a, refuse(request(give(nail))), 2)$
$tell(a, b, challenge(refuse(request(give(nail)))), 3)$
$tell(b, a, justify(refuse(request(give(nail))), \sim have(nail)\}), 4)$

In general, a dialogue is a sequence of *dialogue moves* or *primitives*, where a primitive is represented in the form of an atom in the predicate $tell$. Such predicate has four arguments, respectively: the sender, the recipient, the subject, and the time of the primitive. Time is understood as a *transaction* time, rather than actual time.

Notice that such primitives are tailored to the needs of a two-agent dialogue setting, where, in particular, the recipient is one specific agent. No support for broadcast or multi-cast is provided. However, the framework presented in [13], as described in the next section, is independent of the concrete negotiation language. In Section 5 we will introduce a language supporting multi-casting (via a blackboard) and allowing one-to-many communication primitives. The language that we will propose for auctions is again a set of primitives expressed in the form of $tell$ predicates, with a fundamental difference: the second argument can either be a single agent, or a $group$ of agents. We assume that agents can join groups, uniquely identified in the system. In particular, such identification can be either universal, including *all* agents in the system, or can be used to specify the group composed by the subscribers of a specific auction, as we will see in the following. All of this is in line with the current research on agent systems and interactions, e.g., with KQML. In fact, KQML provides support for multi-casting in a very similar way. A difference with KQML is that in our setting we do not need to introduce predicates other than $tell$. In fact, as we will see later, the (past) dialogue is used by the abductive proof-procedure that generates the dialogue itself as a monotonically growing knowledge base (see for instance the definition of predicate *on-sale* in Section 6). In particular, we do not make use of non-monotonic primitives such as the KQML *untell*, for all utterances keep holding in the utterer's knowledge base.

3 A Negotiation Framework

In addition to the negotiation language, the ingredients needed in building a dialogue framework are: a knowledge representation formalism, a proof procedure for reasoning automatically with the knowledge, and a communication layer. We will not get into details in the description of the knowledge representation and proof procedure on which the agents base their reasoning, and we will mostly refer to [13]. In brief, as far as the knowledge representation is concerned, we will assume that agents have a (declarative) representation of goals, beliefs, and intentions, namely plans to achieve goals. As an example, the knowledge of some agent a can be the following $\mathcal{K}_a$:

$\mathcal{B}_a$ domain-specific beliefs, e.g., $is\_agent(b), i\_am(a)$, as well as domain-independent beliefs, e.g., those implementing negotiation protocols and policies;
$\mathcal{R}_a$: $\{ \, have(picture), have(hammer), have(screwdriver,) \, \}$;
$\mathcal{I}_a$: $\{ \, available(\{hammer, picture\}), plan(\{obtain(nail), hit(nail), hang(picture)\}, 0), goal(\{hung(picture)\}), missing(\{nail\}, 0) \, \}$;
$\mathcal{D}_a$: $\emptyset$.
$\mathcal{G}_a$: $hung(picture)$;

where $\mathcal{B}_a$ stands for agent a's *beliefs*, $\mathcal{R}_a$ stands for *resources*, $\mathcal{I}_a$ for *intentions*, $\mathcal{D}_a$ for (past) *dialogue*, and $\mathcal{G}_a$ for *goals*.

The beliefs can include knowledge that can be used to generate plans. Here we do not make any assumptions on how plans are generated, whether by a planner or from existing libraries.

The beliefs also include *dialogue constraints* that express how agents should react to dialogue moves of other agents. A very simple example of dialogue constraints is the following:

$$tell(X, a, \mathbf{request}(\mathbf{give}(R)), T) \wedge have(R, T) \Rightarrow tell(a, X, \mathbf{accept}(\mathbf{request}(\mathbf{give}(R))), T+1)$$

This constraints reads as follows: 'if agent a receives a request from another agent, X, about a resource R that it has, then a tells X that it will accept the request'[1] A formal specification for $have$ and $need$ is given in [13].

Sets of dialogue constraints express shared protocols and/or individual policies of agents. Dialogues such as the one illustrated in the previous section can be generated automatically within a computational logic setting, e.g. by means of an abductive proof procedure executed within an agent cycle [5]. In computational logic [4], abduction is a reasoning mechanism that allows to find suitable explanations to given observations or goals, based on an abductive logic program. In general, an abductive logic program is a triple $\langle P, \mathcal{A}, IC \rangle$, where P is a logic program, $\mathcal{A}$ is a set of so-called *abducibles* or *hypotheses*, i.e., *open* atoms which can be used to form explanations, and IC is a set of *integrity constraints*, i.e. sentences that need to be satisfied by all explanations. Given a goal g, abduction aims at finding a set of abducibles that, if used to enlarge P, allow to *entail* g while satisfying IC.

The adoption of automatic proof procedures such as that of [1] or [3], supported by a suitable agent cycle such for instance the *observe-think-act* of [5], implements a concrete concept of entailment with respect to knowledge bases expressed in abductive logic programming terms. The execution of the proof procedure within the agent

[1] In our setting, a saying that it accepts X's request about R is equivalent to saying that a *gives* X R: not being concerned with execution, we assume that once an agent tells that it will give a resource, it will actually do it at some point in the future.

cycle allows to produce hypotheses (explanations) that are consistent with the agent integrity constraint, IC. Constraints play a major role in abduction, since they are used to drive the formulation of hypotheses and prevent the procedure from generating wrong explanations to goals.

In [13], abduction has been used to model agent dialogue, with dialogue constraints being represented as integrity constraints, and the beliefs being represented as abductive logic programs. Dialogue constraints are fired each time the agent is expected to produce a dialogue move, e.g., each time another agent sends a request for a resource. Such move is then produced as an hypothesis that must be assumed true in order to keep the knowledge base satisfying the ICs. The use of abduction in the agent dialogue context, as opposed to other (less formal) approaches, has several advantages, among which the possibility to determine properties of the dialogue itself, and the one-to-one relationship holding between specification and implementation, due to the operational semantics of the adopted abductive proof procedure.

4 Auctions

Auctions could be interpreted as an alternative way to negotiate and retrieve resources. They provide an alternative to one-to-one negotiation, e.g. the negotiation carried out by means of dialogues as introduced in the previous section. Indeed, in a framework that includes auctions as a negotiation mechanism, if an agent x needs a resource, it can either ask it to another agent in the system, or find out an auction where the resource is sold and try to obtain it.

Differently from the general dialogue framework, in auctions the *price* of items plays a major role. In particular, the main resource limit is on the initial budget of the single bidders (otherwise agents would buy *at any price*). The participants of an auction are called *auctioneer*, if they sell resources, and *bidders*, if they compete against one another to buy resources. Reasonably, we will assume that there are at least two bidders (otherwise we simply have one-to-one negotiation). In the following, we will call a the auctioneer, $b_1, b_2, \ldots, b_n$ the bidders.

In general, an auction consists of at least three steps, as shown in Figure 1.

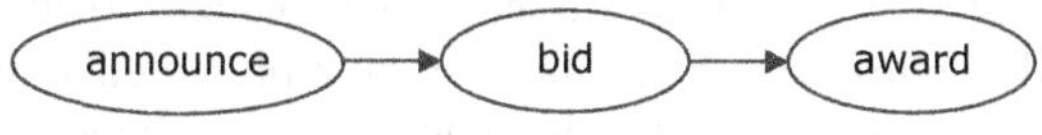

Fig. 1. Basic Steps of an Auction.

The *announcement* step is an utterance made by the auctioneer a, where it declares that an item is on sale at a certain price, for a certain timeout. The *bid* phase is an offer made by a bidder, who is willing to buy the item on sale for a certain price. The *award* phase is a statement made by a to notify everybody that the item has gone[2].

[2] In general, items could also be declared unsold. For the sake of simplicity, we will consider this as a sub-case of this last award phase.

There are various kinds of auctions in literature. We will cite here only four of them, that we consider representative, and in particular: the *English* auction, the *Dutch* auction, the *First Price Sealed Bid (FPSB)* auction, and the *Vickrey* auction.

English Auction. In the case of the English auction, the price of the item that is currently on sale is increased until there is only one bid. The highest bid wins. This behaviour can be represented as in Figure 2, where there can be any number of bids before moving on to the award.

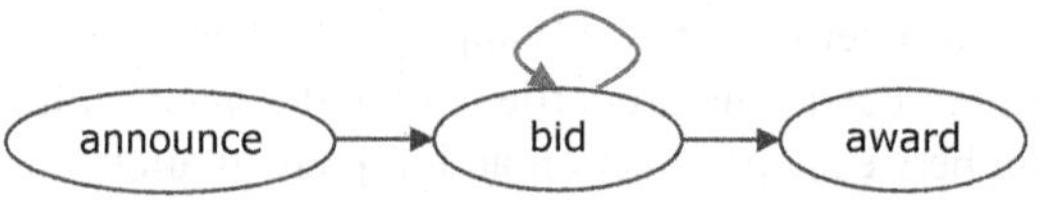

Fig. 2. English Auction: Price Goes Up, 1 Bid at a Time.

Dutch Auction. In the case of the Dutch auction, the price of the item that is currently on sale goes down until there is one bid. The first bid wins. This behaviour can be represented as in Figure 3, where the announcement can be repeated several times before a bid is made.

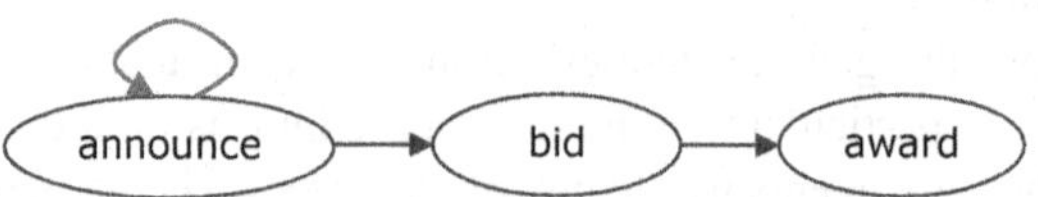

Fig. 3. Dutch Auction: Price Goes Down, 1 Bid at a Time.

FPSB Auction. In the case of the First Price Sealed Bid auction, k bids are simultaneously made by k bidders, $k \leq n$. Bids are sealed, i.e., bidders do not know each other's bids. The highest bid wins. This behaviour can be represented as in Figure 4, where the announcement is made once and k simultaneous bid are made. This protocol is in general faster that the two previous ones, since it is guaranteed to terminate in three steps.

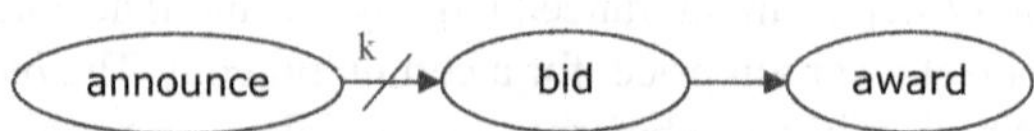

Fig. 4. FPSB Auction: k Bids Are Collected, $k < n + 1$, Highest Wins.

Vickrey Auction. In the case of the Vickrey auction, again, k bids are simultaneously made by k bidders, $k \leq n$. Bids are sealed, i.e., bidders do not know each other's bids. The difference with the FPSB case is that the second highest bid wins. This behaviour is not different from that represented in Figure 4,

It is worth to notice that, by a protocol point of view, Vikrey and FPSB auctions are the same. In fact, the only difference is the final price that the winner pays, which can play a role in terms of strategies, but not in terms of protocols.

Before the auction takes place, the auctioneer must declare *which objects* are about to be sold, *when* the auction is starting, *which kind* of auction it is going to be (English, Dutch, etc.), how long the *timeout* is going to be, and so on. We could call the period of such declarations the *publishing* phase. Then, auctions could be assigned unique *identifiers* (IDs), that will be used by all those who need to refer to the auction.

In general, the auctioneer must know who is participating at the auction. We will therefore assume that bidders *subscribe* to the auction before it starts[3] and form a group. The group is named after the auction ID. After all the objects have been either awarded or declared unsold, the auction is *closed.*

In general, we could represent the four protocols as a particular case of the one represented in Figure 5[4]:

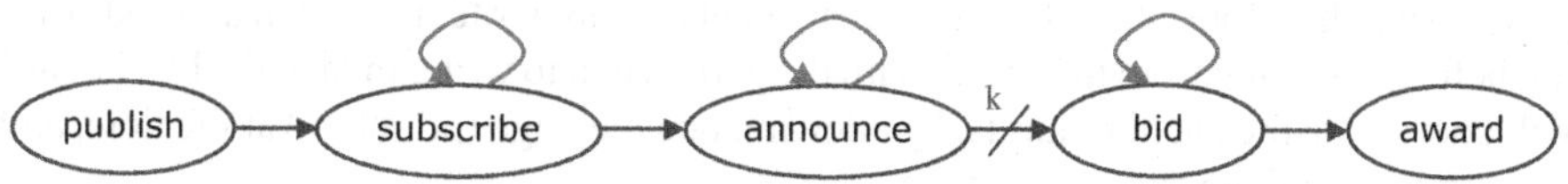

Fig. 5. General Steps of an Auction.

It is clear that it is necessary to have a notion of *timeout*, as the whole auction mechanism is based on it. If we implement auctions by means of dialogue primitives with an associated *transaction* time, we still have to relate it to the *actual* time. To this respect, the blackboard comes in hand, since it can be used to assign primitives both an actual timestamp and a transaction time, and keep them bound somehow. In the sequel, we will assume that agents are equipped with a predicate $actual\_time(Transaction\_time, Actual\_time)$ that keeps track of such relationship.

5 Extended Negotiation Language and Framework

The Coordination Space. If we want to extend this framework to tackle auctions, we need a language, and in addition to it we need a communication layer that provides

[3] More generally, we could assume that bidders can subscribe up until the auction is closed, i.e., up until all the objects have been either awarded or declared unsold.

[4] We omit the closure phase for lack of space. Note that such phase does not play a role, anyway, for the purposes of this paper.

a support for multi-casting. For this we will use a blackboard. Although we will not commit to any particular blackboard or coordination framework, such as for instance a tuple space can be, we will make on the blackboard some assumptions. In particular, we assume that the blackboard is able to assign a transaction time to incoming messages, to store transactions, to manage conflicts in case of multiple concurrently incoming messages (for instance by allowing only one of them), and we will also assume that the blackboard can be programmed in order to allow only those messages that stick to the protocol. For instance, if an English auction is on, where price can only go up, and it must be at least a certain percent higher than that of the previous bid, the blackboard can be programmed to reject those primitives that represent bids that are below the minimum threshold.

We also introduce in the framework a *locking* mechanism, implemented again within the blackboard, that allows bidding only after obtaining the blackboard lock. This is expressed in the agent programs by a logical predicate, $obtain\_lock(\ at(\ Actual\_time\ ),$ $Auction\_ID,\ Transaction\_time\ )$. The semantics of such predicate is: true if the agent obtains the right to write on the blackboard (i.e., to multi-cast) at time $Actual\_time$, the $Transaction\_time$-th primitive of $Auction\_ID$. An example of use of such predicate is in Section 6.

Multi-cast Primitives. As opposed to dialogue primitives, the communication acts of auctions must be multi-casted, which requires a suitable support, as we pointed out previously. Therefore, we will assume that agents do not write directly in a blackboard, not before obtaining from the blackboard the permission to write. In fact, the blackboard will coordinate the agent requests, by allowing one message (primitive) at the time, and by giving it a unique transaction time.

The format of multi-cast primitives must be different from that of one-to-one primitives, since the recipient is in general not unique. As we already said, the second parameter will not be a single agent, but a group of agents. Multi-cast primitives will have in general this format: $tell(Sender,\ Group,\ Subject,\ Time)$, where $Sender$ is an agent, *Group* is a group of agent, intended to be the recipients of the message, $Subject$ expresses the content of the message, and $Time$ is again a transaction time.

We are now ready to introduce a *language* for auctions, i.e., a set of allowed dialogue primitives that agents use to implement an auction:

- $tell(\ Auctioneer,\ all^5,\ publish(\ auction(\ Auction\_ID\ ),\ items(\ \{\ i_1, i_2, \ldots, i_m\ \}\ ),\ protocol(\ english \mid dutch \mid fpsb \mid vickrey\ ),\ beginning\_at(\ initial\_time\ ),\ timeout(\ 2'\ )\ ),\ Time\ )$
- $tell(\ Bidder,\ Auction\_ID,\ subscribe(\ Auction\_ID\ ),\ Time\ )$
- $tell(\ Auctioneer,\ Auction\_ID,\ announce(\ Item,\ Price,\ Timeout\ ),\ Time\ )$
- $tell(\ Bidder,\ Auction\_ID,\ bid(\ Item,\ Price\ ),\ Time\ )$
- $tell(\ Auctioneer,\ Auction\_ID,\ award(\ Item,\ Price,\ Bidder\ ),\ Time\ )$
- $tell(\ Auctioneer,\ Auction\_ID,\ close,\ Time\ )$

[5] Here and in the following, *all* could mean agents all the agents known or reachable by the auctioneer. We assume that the auctioneer has some visibility of the environment; we have not dealt explicitly so far with such issue, though.

Given this language, an example of an English auction is the following [6]:

$\rightsquigarrow$ $tell(\, a, all, publish(\, auction(\, auction_1\,), items(\, \{\, nail\,\}\,), protocol(\, english\,),$
$beginning\_at(\, \texttt{Jan } 3^{\texttt{th}}, 2002, 14:30 \texttt{ GMT}\,), timeout(\, 2'\,)\,), 1\,)$
$\rightsquigarrow$ $tell(\, b_4, a, subscribe(\, auction_1\,), 2\,)$
$\rightsquigarrow$ $\ldots$
$\rightsquigarrow$ $tell(\, a, auction_1, announce(\, nail, 100, 14\,), 13\,)$
$\rightsquigarrow$ $tell(\, b_4, auction_1, bid(\, nail, 100\,), 14\,)$
$\rightsquigarrow$ $tell(\, a, auction_1, announce(\, nail, 110, 16\,), 15\,)$
$\rightsquigarrow$ $tell(\, b_9, auction_1, bid(\, nail, 110\,), 16\,)$
$\rightsquigarrow$ $tell(\, a, auction_1, announce(\, nail, 120, 18\,), 17\,)$
$\rightsquigarrow$ $tell(\, a, auction_1, award(\, nail, 110, b_9\,), 18\,)$ [7]

In this example, the auctioneer starts with publishing the auction $auction_1$ for a certain point in the future ($Jan\ 3^{th}, 2002, 14:30\ GMT$), and collects subscriptions. Then it starts the auction by announcing the first (and only) item, a $nail$. Bidders keep replying to further announcements, until a timeout is reached after the last one. We could give similar examples for the other auctions, but we do not have enough space.

In the case of English and Dutch auctions, as in this example, the auctioneer announces the new (minimum) price and waits for a bid. If the timeout comes the auctioneer posts in the blackboard the award message. We will assume that some aspects, such as filtering unsuitable bids, are managed by the blackboard (bids for items not being sold in that auction, bids from unsubscribed agents, bids for a price that is too low, ... will be rejected by the blackboard). We admit that this is a strong assumption on the environment, and needs to be further investigated. Moreover, such behaviour hardwired in the blackboard, could require a dynamic (re-)configuration. It is not obvious at all who and how should to it. Another important issue, with this respect, is how to merge such non-logical element within the logical framework.

We can assume of course that bidders can bid more than announced by the auctioneer (highest bid wins anyway). This is up to the agent. In the case of FPSB and Vickrey auctions, that we could not represent here, the timeout would be expressed in terms of transaction time, as the current time $+n$, n being the number of expected bidders. This is one more reason why it is required to subscribe an auction: n is actually the number of subscribers. This could raise some problems, in that it would require some assumptions on the presence of all subscribers for the whole duration of the auction, which is maybe too strong in a multi-agent setting. Anyway, it is possible to relax this condition, and define a protocol in which not all subscribers must bid. We could imagine that if there are less bids than bidders ($k < n$), the auctioneer will be in charge to make sure that before too late (*actual* timeout) n messages are anyway posted in the blackboard, in order to reach the (*transaction*) timeout. She will therefore put $n - k$ messages atomically together with the award message, just before it is published.

[6] See Appendix for more examples

[7] As we said before, we will not deal with the closure of auctions, which could be expressed by a primitive such as $tell(\, a, auction_1, close, 19\,)$, uttered by the auctioneer.

6 An Example: Implementation of the English Auction Protocol

In this section, we will give here an example of a possible implementation of the English auction protocol. Due to the space constraints, we will only write the basic rules and constraints of the two kinds of agents: the bidder and the auctioneer. Let us start with the auctioneer program.

auctioneer (a) program:

$$\begin{aligned}&\mathbf{on\_sale(Item, Auction\_ID)} \leftarrow\\ &\quad tell(a, all, publish(auction(Auction\_ID), items(Items),\\ &\quad protocol(Protocol), beginning\_at(Initial\_time), timeout(Timeout)), Time)\\ &\quad \wedge\, Item \in Items\\ &\quad \wedge\, actual\_time(Actual\_time)\\ &\quad \wedge\, Initial\_time < Actual\_time\\ &\quad \wedge\, not\ timeout\_expired(Item, Auction\_ID, Timeout)\\ &\quad \wedge\, not\ (tell(a, Auction\_ID, award(Item, Price, Bidder), Time1)\\ &\qquad \wedge\, Time1 < Time)\end{aligned}$$

An item is *on_sale* if it is included in the item list of an ongoing auction, and the timeout after the last bid has not expired.

$$\begin{aligned}&\mathbf{timeout\_expired(Item, Auction\_ID, Timeout)} \leftarrow\\ &\quad tell(a, Auction\_ID, announce(Item, Price1, Timeout1), Time1)\\ &\quad \wedge\, not\ (tell(a, Auction\_ID, announce(Item, Price2, Timeout2), Time2)\\ &\qquad \wedge\, Time2 > Time1)\\ &\quad \wedge\, actual\_time(Time1, Actual\_time1)\\ &\quad \wedge\, actual\_time(now, Actual\_time)\\ &\quad \wedge\, Actual\_time > Actual\_time1 + Timeout\end{aligned}$$

The timeout that refers to a certain item on sale at a certain auction is *expired* if *now* is a later time than that of the last announcement, plus the declared $Timeout$.

$$\mathbf{new\_price(Price, New\_price)} \leftarrow New\_price\ is\ Price + 0.1 * Price$$

The auctioneer announces the new price of an item adding a ten percent to the last price.

$$\begin{aligned}&\mathbf{tell(Bidder, Auction\_ID, bid(Item, Price), Time)}\\ &\wedge\, on\_sale(Item, Auction\_ID)\\ &\wedge\, new\_price(Price, New\_price)\\ &\wedge\, obtain\_lock(at(now), Auction\_ID, Time + 1)\\ &\quad \Rightarrow \mathbf{tell(a, Auction\_ID, announce(Item, New\_price, Time + 2), Time + 1)}\end{aligned}$$

This constraints says that if a participant bids at time $Time$, and the item in question is still on sale, the auctioneer next (i.e., at $Time + 1$) announces such item for a new price. Such announcement will hold up to $Time + 2$.

$\mathbf{timeout\_expired(Item, Auction\_ID, Timeout)}$
$\wedge\ tell(A, all, publish(auction(Auction\_ID), \ldots, timeout(Timeout)), \ldots)$
$\wedge\ tell(Bidder, Auction\_ID, bid(Item, Bidder\_price), Time - 1)$
$\wedge\ tell(A, Auction\_ID, announce(Item, New\_price, Time + 1), Time)$
$\wedge\ Bidder\_price \geq New\_price$
$\quad \Rightarrow \mathbf{tell(a, Auction\_ID, award(Item, Bidder\_price, Bidder), Time + 1)}$

This constraints says that after the timeout has expired the highest bid wins, and therefore the bidder is awarded the item.

$\mathbf{timeout\_expired(Item, Auction\_ID, Timeout)}$
$\wedge\ tell(A, all, publish(auction(Auction\_ID), \ldots, timeout(Timeout)), \_)$
$\wedge\ tell(A, Auction\_ID, announce(Item, \_, \_), Time)$
$\wedge\ not\ tell(\_, Auction\_ID, bid(Item, \_), \_)$
$\quad \Rightarrow \mathbf{tell(a, Auction\_ID, award(Item, \ldots, noone), Time + 1)}$

This constraints says that if there are no bids, the item is declared unsold.

bidder (*b*) program:

The bidder program is simpler, but it introduces two predicates whose implementation is not specified here, and will depend on the individual bidder programs (we assume that each bidder may adopt a different policy from the others). Such predicates are $calculate\_new\_price$ and $calculate\_bid\_time$.

$\mathbf{tell(a, Auction\_ID, announce(Item, Price, Timeout), Time)}$
$\wedge\ have\_subscribed(Auction\_ID)$
$\wedge\ not\ (tell(Participant, Auction\_ID, Anything, Time1) \wedge Time1 > Time)$
$\wedge\ calculate\_new\_price(Item, Price, New\_price)$
$\wedge\ calculate\_bid\_time(Auction\_ID, Bid\_time)$
$\wedge\ obtain\_lock(at(Bid\_time), Auction\_ID, Time + 1)$
$\quad \Rightarrow \mathbf{tell(b, Auction\_ID, bid(Item, New\_price), Time + 1)}$

This means that if the auctioneer has announced an $Item$ at a $Price$, at the (transaction) time $Time$, and after that no other participant made any move, if the bidder b can calculate a suitable price and can obtain the blackboard to bid at a certain time (Bid_time), then b will bid at (transaction time) $Time + 1$. The two above mentioned predicates and the blackboard will decide whether the agent will actually bid or not.

$\mathbf{have\_subscribed(Auction\_ID)} \leftarrow$
$\quad tell(b, a, subscribe(Auction\_ID), Time)$

7 Discussion and Future Work

We have extended an existing framework for one-to-one agent negotiation to cope with one-to-many negotiation, in the particular case of auctions. There is much ongoing work on the negotiation area: those interested could refer to [11] for an introduction, and to [6] and [2] for a perspective on state of the art methods and challenges.

An other existing approach to negotiation via dialogue, that makes use of an argumentation system, is that of [15], and more recently [7], where the authors adopt a modal logic approach and focus on the mental states of agents that move towards an agreement, and on the way to persuade a counterpart in order to foster cooperation.

As far as auctions, there is much work on mathematical models, and protocol comparison in terms of efficiency, stability, etc., but to the best of our knowledge, only a little relate theoretically founded models to operational models, as in our case.

This is a preliminary work on the subject and needs to be further expanded in several ways. The main issue, in our opinion, is how to tackle the various non-logical elements, such as the initial time, the timeout, the blackboard lock, and merge them in the logical framework that we propose. This is a need because auctions are intrinsically non-logical in relying on the concept of timeout. As opposed to the case of dialogue primitives, in an auction setting the trigger that fires a dialogue constraint and makes a primitive being cast will necessarily depend not only on the existence of other incoming messages, but also on the *lack* of such messages. That is, on the fact that a timeout is met.

However, in our framework, we manage to separate such non-logical features from the logical reasoning bit, partly relegating them to the blackboard, partly implementing them inside logical predicates used by the agent.

As future work, we would like to integrate within this framework the two negotiation patterns, giving the agents the possibility to either retrieve the missing resources by means of dialogues, or by means of auctions. We aim at defining a comprehensive architecture in which an agent can choose to obtain a resource either through an auction or through a 'simple' sequence of dialogues.

Another direction of research is that of *policies*. We consider it important to provide the possibility to define policies that are orthogonal to the agent program used to multicast primitives, and to the constraints in particular. For instance, before committing to a bid, agents could reason in terms of available budget, need for other items, maximum price that they intend to pay for the item, importance of the goal, attitude they may have with respect to the other bidders, etc. Together with policies, *strategies* can be used to maximize an agent's payoff, e.g. agents can decide to bid as soon as possible, or as late as possible, with the adopted policy.

We intend to study properties of auctions, when different policies and strategies meet. Finally, we plan to give an implementation of this one-to-many framework, possibly in integration with the other dialogue-based negotiation framework.

Acknowledgements

We would like to thank Paola Mello, Maurelio Boari, and the anonymous referees, for their precious comments, suggestions, and support.

References

1. T. H. Fung and R. A. Kowalski. The iff proof procedure for abductive logic programming. *Journal of Logic Programming*, 1997.
2. N. R. Jennings, P. Faratin, A.R. Lomuscio, S. Parsons, C. Sierra, and M. Wooldridge. Automated negotiation: Prospects, methods and challenges. *International Journal of Group Decision and Negotiation*, 10(2), 2001.
3. A. Kakas and P. Mancarella. On the relation between truth maintenance and abduction. In T. Fukumura, editor, *Proceedings of the first Pacific Rim International Conference on Artificial Intelligence, PRICAI-90, Nagoya, Japan*, pages 438–443, 1990.
4. A. C. Kakas, R. A. Kowalski, and F. Toni. The role of abduction in logic programming. *Handbook of Logic in AI and Logic Programming*, 5:235–324, 1998.
5. R. A. Kowalski and F. Sadri. From logic programming to multi-agent systems. *Annals of Mathematics and AI*, 1999.
6. S. Kraus. *Strategic Negotiation in Multi-Agent Environments*. MIT Press, Cambridge, MA, 2000.
7. S. Kraus, K. Sycara, and A. Evenchik. Reaching agreements through argumentation; a logical model and implementation. *Artificial Intelligence*, 104:1–69, 1998.
8. S. Parsons, C. Sierra, and N. R. Jennings. Agents that reason and negotiate by arguing. *Journal of Logic and Computation*, 8(3):261–292, 1998.
9. A. Rao and M. Georgeff. An abstract architecture for rational agents. In *Proceedings of the International Workshop on Knowledge Representation (KR'92)*, 1992.
10. S. Rochefort, F. Sadri, and F. Toni, editors. *Proc. International Workshop on Multi-Agent Systems in Logic Programming, in conjunction with ICLP'99, Las Cruces, New Mexico.* November 1999.
11. J. S. Rosenschein and G. Zlotkin. *Rules of Encounter: Designing Conventions for Automated Negotiation Among Computers*. MIT Press, Cambridge, Massachusetts, 1994.
12. F. Sadri, F. Toni, and P. Torroni. Dialogues for negotiation: agent varieties and dialogue sequences. In *Proceedings ATAL'01, best paper award, Seattle, WA*, August 2001.
13. F. Sadri, F. Toni, and P. Torroni. Logic agents, dialogues and negotiation: an abductive approach. In *Proceedings AISB'01 Convention, York, UK*, March 2001.
14. K. Satoh and F. Sadri, editors. *Proc. Workshop on Computational Logic in Multi-Agent Systems (CLIMA-00), in conjunction with CL-2000, London, UK.* July 2000.
15. K. P. Sycara. Argumentation: Planning other agents' plans. In *Proceedings 11th International Joint-Conference on Artificial Intelligence*, pages 517–523. Morgan Kaufman, 1989.
16. G. Weiss. *Multiagent Systems*. MIT Press, 1999.

Appendix: More Examples Of Automated Auction: Dutch, Fpsb, and Vickrey Auctions

An example of a Dutch auction is the following:

$\rightsquigarrow$ $tell(\ a, all, publish(\ auction(\ auction_2\ ), items(\ \{\ nail\ \}\ ), protocol(\ dutch\ ),$
$beginning\_at(\ \texttt{Jan}\ 3^{\texttt{th}}, 2002, 14:30\ \texttt{GMT}\ ), timeout(\ 2'\ )\ ), 1\ )$
$\rightsquigarrow$ $tell(\ b_4, a, subscribe(\ auction_2\ ), 2\ )$
$\rightsquigarrow$ ...
$\rightsquigarrow$ $tell(\ a, auction_2, announce(\ nail, 100, 3\ ), 12\ )$
$\rightsquigarrow$ $tell(\ a, auction_2, announce(\ nail, 90, 4\ ), 13\ )$
$\rightsquigarrow$ $tell(\ a, auction_2, announce(\ nail, 80, 5\ ), 14\ )$
$\rightsquigarrow$ $tell(\ a, auction_2, announce(\ nail, 75, 6\ ), 15\ )$
$\rightsquigarrow$ $tell(\ b_3, auction_2, bid(\ nail, 75\ ), 16\ )$
$\rightsquigarrow$ $tell(\ a, auction_2, award(\ nail, 75, b_3\ ), 17\ )$

An example of a FPSB auction is the following (the price is encrypted; example with 5 bidders):

$\rightsquigarrow$ $tell(\ a, all, publish(\ auction(\ auction_3\ ), items(\ \{\ nail\ \}\ ), protocol(\ fpsb\ ),$
$beginning\_at(\ \texttt{Jan}\ 3^{\texttt{th}}, 2002, 14:30\ \texttt{GMT}\ ), timeout(\ 2'\ )\ ), 1\ )$
$\rightsquigarrow$ $tell(\ b_4, a, subscribe(\ auction_3\ ), 2\ )$
$\rightsquigarrow$ ...
$\rightsquigarrow$ $tell(\ a, auction_3, announce(\ nail, 100, 12\ ), 7\ )$
$\rightsquigarrow$ $tell(\ b_1, auction_3, bid(\ nail, \$105\$\ ), 8\ )$
$\rightsquigarrow$ $tell(\ b_5, auction_3, bid(\ nail, \$103\$\ ), 9\ )$
$\rightsquigarrow$ $tell(\ b_3, auction_3, bid(\ nail, \$104\$\ ), 10\ )$
$\rightsquigarrow$ $tell(\ b_2, auction_3, bid(\ nail, \$106\$\ ), 11\ )$
$\rightsquigarrow$ $tell(\ b_4, auction_3, bid(\ nail, \$101\$\ ), 12\ )$
$\rightsquigarrow$ $tell(\ a, auction_3, award(\ nail, \$106\$, b_2\ ), 13\ )$

An example of a Vickrey auction is the following (same bids as before):

$\rightsquigarrow$ $tell(\ a, all, publish(\ auction(\ auction_4\ ), items(\ \{\ nail\ \}\ ), protocol(\ vickrey\ ),$
$beginning\_at(\ \texttt{Jan}\ 3^{\texttt{th}}, 2002, 14:30\ \texttt{GMT}\ ), timeout(\ 2'\ )\ ), 1\ )$
$\rightsquigarrow$ $tell(\ b_4, a, subscribe(\ auction_3\ ), 2\ )$
$\rightsquigarrow$ ...
$\rightsquigarrow$ $tell(\ a, auction_4, announce(\ nail, 100, 12\ ), 7\ )$
$\rightsquigarrow$ $tell(\ b_1, auction_4, bid(\ nail, \$105\$\ ), 8\ )$
$\rightsquigarrow$ ...
$\rightsquigarrow$ $tell(\ b_2, auction_4, bid(\ nail, \$106\$\ ), 11\ )$
$\rightsquigarrow$ ...
$\rightsquigarrow$ $tell(\ a, auction_4, award(\ nail, \$105\$, b_13\ ), 8\ )$

The Tragedy of the Commons - Arms Race within Peer-to-Peer Tools

Bengt Carlsson

Blekinge Institute of Technology
371 25 Ronneby, Sweden
bengt.carlsson@bth.se

Abstract. The two major concerns about peer-to-peer are anonymity and non-censorship of documents. Music industry has highlighted these questions by forcing Napster to filter out copyright protected MP3 files and taking legal actions against local users by monitoring their stored MP3 files. Our investigation shows that when copyright protected files are filtered out, users stop downloading public music as well. The success of a distributed peer-to-peer system is dependent on both cooperating coalitions and an antagonistic arms race. An individual will benefit from cooperation if it is possible to identify other users and the cost for doing services is negligible. An arms race between antagonistic participants using more and more refined agents is a plausible outcome. Instead of "the tragedy of the common" we are witnessing "the tragedy of arms race within the common". Arms race does not need to be a tragedy because these new tools developed or actions taken against too selfish agents may improve the P2P society.

1 Background

The Internet as originally conceived back in the late 1960s was fundamentally designed as a peer-to-peer (P2P) system (see [7] for a historical overview). The early Internet was more open and free than today's network. Generally any two machines could send packets to each other. Early client/server applications like FTP and Telnet had a symmetric usage pattern. Every host on the Internet could FTP or Telnet to any other host with servers usually acting as clients as well.

The Domain Name System (DNS), originally from the early 1980s, is an example of a system that blends P2P networking with a hierarchical model of information ownership. A DNS queries higher authorities about unknown names getting answers as well as new queries back. Name servers operate both as clients and as servers.

The explosion of the Internet in 1994 radically changed the shape of Internet into a commercial mass cultural phenomenon. People were connected to the Internet using modems, companies installed firewalls, and the initial structures broke down.

Uncooperative people use the Internet in their own interest without looking at the interests of the common. In the first half of the 1990s this was a surprising experience on the Internet. The "Green Card spam" 1994 appeared on the Usenet as an advertisement posted individually to every Usenet newsgroup. The advertisers did not

A. Omicini, P. Petta, and R. Tolksdorf (Eds.): ESAW 2001, LNAI 2203, pp. 119-133, 2001.

pay for the transmission of the advertisement; the costs were born by the Usenet as a whole. Today we have to assume that users behave selfishly and/or have commercial interests. This is a fundamental transition of the Internet and the main topic of this article.

With slow speed modem connections (and large phone bills), user patterns normally involved downloading data, not publishing or uploading information. Companies on the other hand hid their data behind firewalls making it hard to upload data from outside the firewall. By default, any host that can access the Internet can also be accessed on the Internet. Behind the firewalls this was no longer true making the need for a permanent IP address unnecessary for the end-user when IP addresses became in short supply. With dynamic IP addresses the single user is hard to find outside the local network.

Users were getting better computer performance and more applications, but with less authorities than in the early days of the computers. In the late 1990s programs started to bypass DNS in favor of creating independent directories of protocol-specific addresses. Examples are ICQ and Napster, the latter a tremendous success with over 80 million non-DNS addresses in less than two years.

A suggested litmus test that determines whether a system is P2P or not is suggested by [10]. If the answer to both questions below is yes, the application is P2P. If the answer to either question is no, it's not P2P:

1. Does it treat variable connectivity and temporary network addresses as the norm?
2. Does it give the nodes at the edges of the network significant autonomy?

P2P systems can be classified [6] into three main categories: hierarchical, centrally coordinated, and decentralized.

1. A hierarchical P2P system organizes peers into hierarchies of groups where communication is coordinated locally or passed upwards to a higher-level coordinator for peers communicating between groups. A DNS fulfills the requirements of the second question but uses permanent IP-addresses
2. In a centrally coordinated system, coordination between peers is controlled and mediated by a central server. SETI@home is a project trying to detect intelligent life outside earth, which distributes necessary data to millions of end-user computers during screen saving periods or as a process. Because of the biased information flow there is little autonomy left for the end-users, i.e. the significant autonomy criterion in the second question is not fulfilled. Napster stores pointers and resolves addresses of MP3 files and users centrally, but leaves the contents and sharing of the files at the users' machines. This is a true P2P system.
3. Completely decentralized P2P systems have no notion of global coordination at all. Communication is handled entirely by peers operating at a local level, where messages may be forwarded on behalf of other peers. Freenet is mainly focusing on preventing censorship of documents and providing anonymity for users on the Internet. An example is Gnutella which has been used for distributing MP3 music as well as picture and video files among end users.

We will examine the present development of centrally coordinated and decentralized P2P distribution of MP3 files within Internet using models taken from evolutionary biology. In section 2 the concept of the tragedy of the commons is discussed, followed by a description of the different P2P systems investigated. The research results are described in section 4. P2P systems are further discussed in section 5, and finally some concluding remarks are made in section 6.

2 The Tragedy of the Commons

In the background section we have seen the transformation of the Internet from a typical cooperative platform to a highly competitive network. Deeper studies of the concept of human nature is outside the scope of this article but let us make some comprehensive statements about humans as biological beings and part of natural ecosystems.

Garret Hardin [5], using a game theoretic model of explanation, described the conflict between the individual and the common in "the tragedy of the commons" as follows:

"Ruin is the destination toward which all men rush, each pursuing his own best interest in a society that believes in the freedom of the commons. Freedom in a commons brings ruin to all."

Within the P2P field this metaphor is used by several researchers (e.g. [1,8] for explaining overexploitation of the resources.

So are we destined for being either definite controlled or vulnerable to a chaotic Internet? To answer this question we may look at other distributed, open systems capable of evolving robust behaviors based on autonomous selfish agents. Robustness is the balance between efficiency and efficacy necessary for survival in many different environments. A bio ecosystem exactly corresponds to these conditions. The main principle of a biological ecosystem is natural selection [12,13]. This selection, the survival of the fittest, happens among individuals, or agents, with opposed competing skills. A dynamic process, where the action of one agent is retorted by counter-actions taken by another agent, is starting an arms race. In the end one group of agents may form coalitions against another group, based on the needs of the individuals but dependent upon the success of the coalition (see also [3]).

Instead of looking for a central coordinator, the ecosystem emerges by the dynamics of the autonomous agents. This vision is shared by the recent FET (Future and Emerging Technologies) initiative "Universal Information System (UIE)" within the Information Society Technologies (IST) program of the European Commission[1] and by Internet Ecologies Area's[2] which focuses on the relation between the local actions and the global behavior of large distributed systems, both social and computational.

The global information infrastructure may be regarded as an emerging information ecosystem of infohabitants, or agents. Within information ecosystems infohabitants who may have opposite interests, perform the activities. This dynamic process may be compared to a biological view of describing ecosystems, where skills and interactions determine the success of the infohabitants. A biological system does explain the advantage of having cooperating agents within well performing ecosystems, by its intrinsic dynamics. Such a robust ecosystem will eliminate the advantage for infohabitants of being too disloyal against the community.

Multi-agent system (MAS) has so far paid little attention to the ideas surrounding P2P computing although agent techniques have been applied to the design and implementation of interesting decentralized applications. It is intuitive to think of

---

1 http://www.cordis.lu/ist/fetuie.htm.

2 http://www.parc.xerox.com/istl/groups/iea/ .

MAS as P2P systems, since many agents and/or hosts in MAS have been thought of as networks of equal peers. Similarly many existing P2P system can be thought of in terms of concepts developed by the MAS community, e.g., Napster can be thought of as a matchmaker

In our investigation, agents (and the users behind) are autonomous and selfish. Instead of focusing on normative agents, the emphasis is on the dynamics of a competitive system. The tragedy of the commons includes autonomous and selfish agents violating norms and starting an arms race.

3 Four Different Peer-to-Peer Systems for File Sharing

We investigated four different P2P tools that range from centrally coordinated to completely decentralized systems. Napster and MusicCity represent centrally coordinated system while BearShare represents a decentralized system. The fourth tool, CatNap, encrypts MP3 files in order to bypass the filtering function of Napster.

3.1 Napster

Napster is connecting users using file registers. MP3 files are stored in the computers of the users, but Napster keeps a track of all the filenames. Originally Napster did not separate free and copyright protected music, which made the Recording Industry Association of America (RIAA) take legislative counter-measures. After a court order Napster must provide for songs to be blocked, by filtering out all copyright protected songs from their file register. Downloaded music files both successful and interrupted are locally logged together with IP addresses in this investigation.

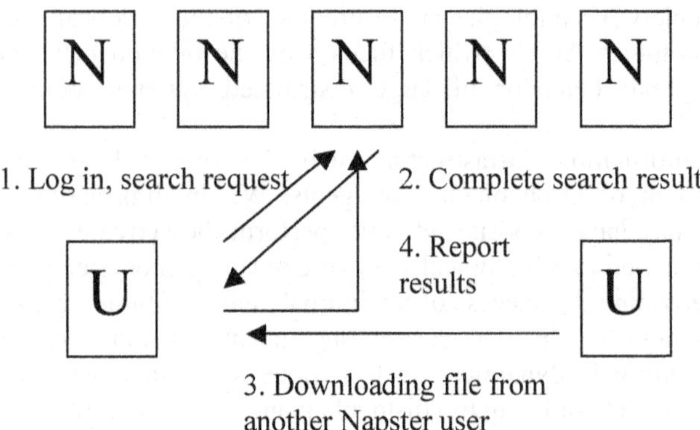

Fig. 1. Downloading files using Napster where N is the Napster server and U denotes the user.

Using Napster for finding MP3 files is basically a client – service task finding the requested file and a client – client task transferring the file. As can be seen in Fig. 1 the protocol for doing this in principle involves only four steps. The user is doing a login and search request followed by a search result answer from the Napster server. The file is downloaded from another user and the result is reported back to the Napster server. It is possible for the Napster user to stop other users from downloading the locally stored files, i.e. a user may be selfishly.

3.2 BearShare

BearShare, a Gnutella client, is a distributed P2P tool connecting to at most seven other peers. It has a monitoring tool, which registers the download as before but also downloading time and original country for IP addresses.

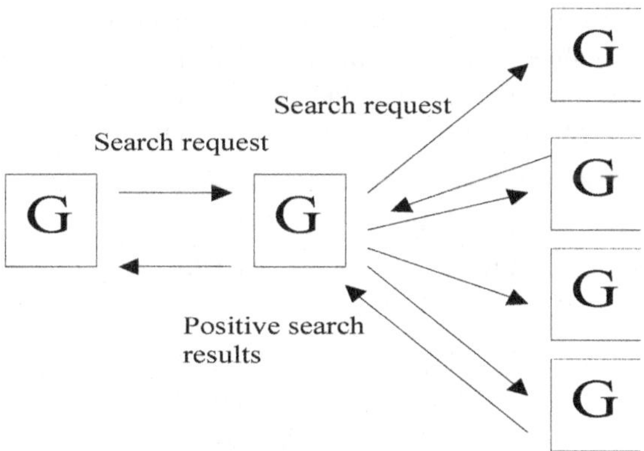

Fig. 2. Downloading files using Gnutella where G is the Gnutella server. Only a small fraction of the actual search space is shown.

A Gnutella P2P tool uses a huge amount of communication compared to e.g. Napster. After an initial connection to a known Gnutella server a network of Gnutella servers are established. Unlike Napster there is no division into clients and servers at any stage of the service. A representative network includes connecting to four other servers and seven steps of message duration (time to live, TTL equal to 7). In all a server connects to over 4000 other servers.

In Fig. 2 the original Gnutella user is sending a search request to four other users (only one is shown in Fig. 2), which in turn send the request to four other users. With TTL = 7 the message passes five more steps. It is possible to broadcast across firewalls or to drop messages when connected to low-bandwidth networks. Positive search results are sent back to the requester. The user is allowed to alter the settings of the protocol by changing the number of servers connected or changing the value of TTL. A remote P2P tool distributor may use agents inside the program to monitor and distribute the activities of a local peer. This so-called spyware is actually used by BearShare. A selfish user may enlarge the search area and prevent uploading local files. This subject will be further analyzed in the discussion section.

Gnutella may be seen as an information ecosystem of agents. Interactions among agents in both natural and information ecosystems may be regarded as a network of dynamically connected agents. In a small world model [11] the ecosystem is represented as a graph with edges connecting different vertices. The amount of interaction within the ecosystem is dependent on the amount of clustering between agents and the path length for reaching an arbitrary agent. Neither a graph with few neighbor connections nor a fully connected graph will do, because of high path length and low clustering respectively. A small world graph with the combination of high local clustering and short global path lengths will do better. Recently Albert et al [2] have shown that two randomly chosen documents on the web are on average 19 clicks away from each other

3.3 MusicCity

Like BearShare but unlike Napster, MusicCity does not maintain a central content index and is not currently subject to content filtering. A file may be downloaded using more than one source file. Like Napster but unlike BearShare, MusicCity is formally a closed system, requiring centralized user registration and logon.

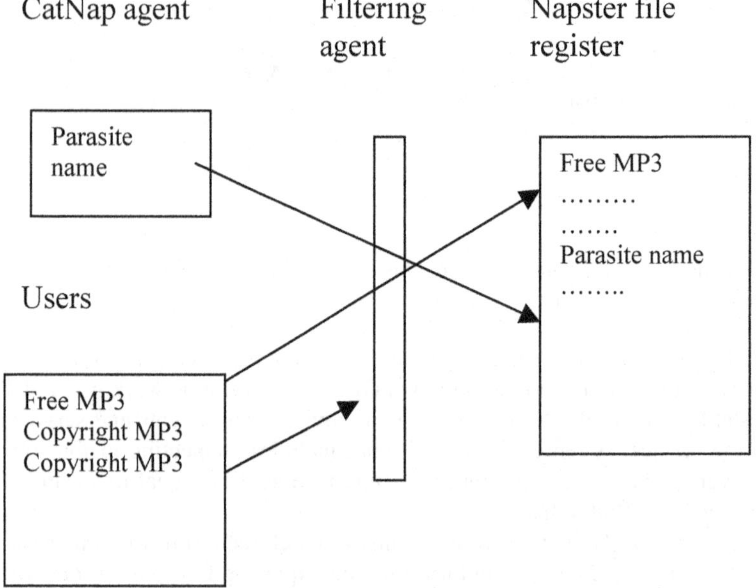

Fig. 3. The Filtering Function of Napster Affected by a Parasite Agent.

3.4 CatNap

We also investigated CatNap, a program working within Napster encrypting MP3 files. CatNap users must convert their files before entering Napster in an attempt to fool the Napster filter. A user inside the CatNap mode has no possibility to participate in the ordinary Napster community, s/he will act as a parasite agent.

Filtering agents and encrypting agents supplement the user-controlled behavior Fig. 3 shows the essential agent interaction. The goal for the filtering agent is to stop copyright protected MP3 filenames to enter the file register. If other free MP3 files are also stopped, the filter has become too efficient. The CatNap files may pass the filter by encrypting the file names. We shall in the investigation part compare the filtered Napster society with a similar unfiltered one.

4 Empirical Study

Three identically equipped computers were used during March to May 2001 to investigate Napster, MusicCity and BearShare. During the period MusicCity changed locations making it impossible to reach during several weeks. Also, there was no measurement available for the downloading via BearShare in the first half of the investigation period.

Napster and MusicCity both have centralized distribution of connecting different peers. This fact makes traffic analysis somewhat unnecessary at the end-user level, because much better predictions are done at central servers. BearShare's distributed propagation makes a local investigation necessary because only directly connected peers are involved in the file sharing. We concentrate our efforts on measuring the substantial contents instead of measuring the actual traffic. With Napster the network consists of 5.000 – 10.000 users (out of a population exceeding 1.000.000) and with MusicCity 15.000 – 35.000 users constitute the total number of logged in users for MusicCity.

The investigation of CatNap was done during a ten-day period, at the end of March 2001. Later on, Napster banned CatNap and other encryption programs by identifying encrypted files and blocking these users from Napster.

4.1 Average Number of Files

When Napster started to use filtering tools, the average number of files made available per user not surprisingly went down. In Fig. 4 bars represent this where the initial, non-filtered files amounted to about 200 per user. At the end of the investigation less than 50 files are left.

The measurring lasted for a month and a half, so every single day is not represented in the diagram. During the first month the efficiency of the filter varied due to a leaking filter. This is shown by the varying number of files in the middle section of the diagram.

During a ten-day period when the Napster filter was still leaking the number of files per user was compared for Napster, MusicCity and CatNap as can be seen in Fig. 5. Roughly MusicCity users twice the number of Napster users and CatNap twice the number of MusicCity users. CatNap files were measured by manually counting the number of files per user (totally 125 users). The MusicCity number of files per user is about the same Napster had before the discussion of filtering files started.

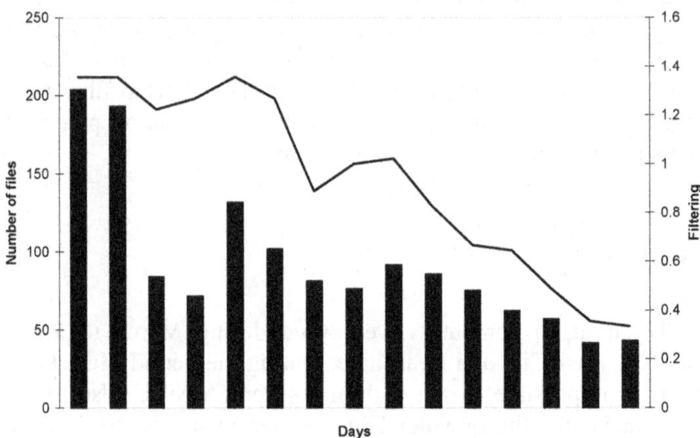

Fig. 4. Mean Number of User Files and Filtering Efficiency within Napster.

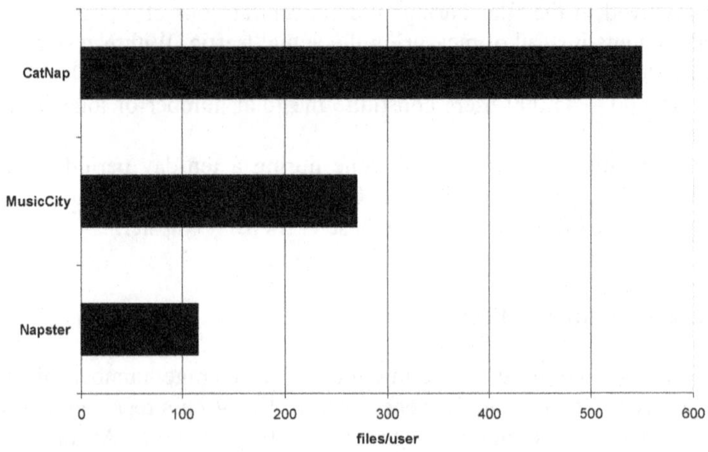

Fig. 5 Files per User for Napster, MusicCity and CatNap.

A possible alternative explanation of the decreasing number of files would be that users with a lot of MP3 files were leaving Napster because of difficulties finding new songs for downloading. To measure this a fixed number (n=61) of both public and copyright protected songs were monitored on a local site. The number of songs passing the filter was measured and an estimated number of songs supposed not to be filtered out are indicated as a filtering efficiency scale in Fig. 4. A filtering effiency close to 1 indicates a well functioning filter neither failing to filter out songs nor overreacting. As can be seen in Fig. 4 the filtering capacity pretty well follows the variation of the number of files. There is no indication of changing file-sharing behavior because of the filter.

4.2 Filtering Efficiency

In Fig. 6 we investigated the proportion of files not filtered out by Napster (15 out of 61 files) compared to the filtered files. These filtered or public files are represented as the white parts of the bars for Napster, MusicCity and BearShare.

The investigation was done during on average 77 hours each period. After a peak in the beginning of April Napster almost disappeared. This decline expresses both the loss of filtered files and a decreased interest for remaining public files (a decrease to 1/8 of the public files peak value). MusicCity users had a 31 % and BearShare a 28 % increase of their downloading rate between the measurements, but no increase at all for Napster´s public files. Users are downloading proportionally more copyright protected files from BearShare and MusicCity.

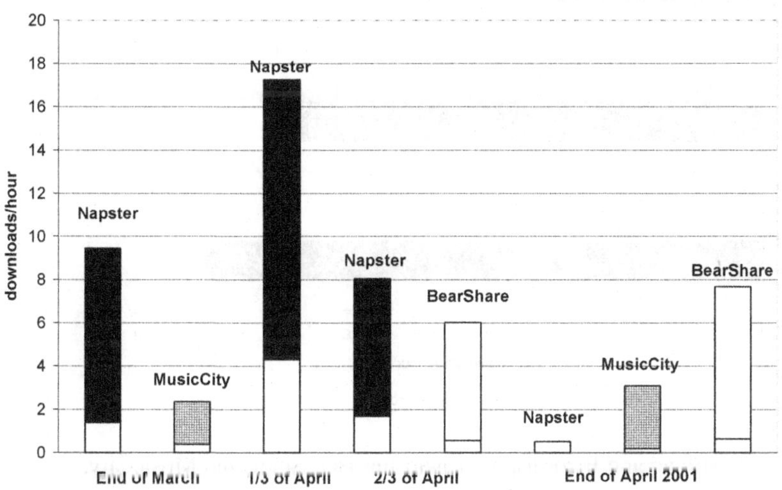

Fig. 6. Average number of downloads per hour for Napster, MusicCity and BearShare. White part of the bar represents files not filtered out by Napster.

4.3 Error Rate

Next we compared the rate of errors when trying to download files. The measurements were done during a ten days period lasting for on average 125 hours for each tool.

Fig. 7 shows that Napster has the lowest rate of errors. This is probably because Napster only allows one-to-one connections when uploading a file. MusicCity allows multiple sources and BearShare distributes a user´s uploads based on the capacity of the remote connection. Because of BearShare´s increased use of communication between peers the download rates are probably near the capacity limit for this connection.

On average it took 20 minutes and 32 seconds (average for 450 files) for downloading a file using BearShare with variations from several hours to a few seconds. For each file there was a mean transfer speed of 48 kbit/s. Napster and most of MusicCity downloading time for files were directly reflecting the speed of the remote connection.

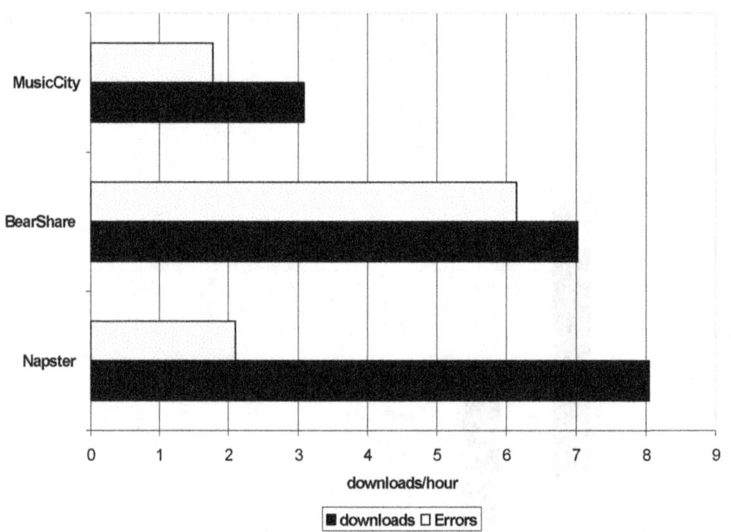

Fig. 7. Download and Error Rate for Napster, BearShare and MusicCity.

4.4 User Reactions

Users do react against Napster´s filtering function by misspelling artist names and songs. This is possible to do because it is the users naming both songs and artists for a certain MP3 file Fig. 8 below shows an example of that. Searching for artist name or full title returned no results, but searching for "Oops" did. Besides finding misspellings of the actual file, all other files withholding "Oops" will be found.

My Files\Britney_Speares-Oops...I_did_it_again.mp3
My Files\weird al yankavitch- oops, i farted again.mp3
My Files\Britery - Oops I did it Again.mp3
MUSIK\Max Raabe - Oops I did it again.mp3
My Files\Oops I Farted Again.mp3
My Files\N'Stink - Oops I Shitted Again.mp3
Palich\oops1.mp3
My Files\Oops! I Did It Again.mp3
My Files\Oops I Did It Again.mp3
Max Raabe - oops-i-did-it-again.mp3
New Folder\Oops! ...I Did It Again.mp3
PHILIP's\oops i did it agn.MP3
My Files\SNL - Oops I counted Again (Al Gore Florida Parody).mp3
My Files\Oops I fucked the real Slim Shaddy again .wma
My Files\Emeniem-(Parody)-Oops I did it Again.mp3
downloads\Oops I did it Again.mp3
My Files\SNL - Al Gore - Oops I Counted Again .mp3
My Files\Chipmunks - Oops! I Did It Again.mp3
Bjork\Bjork & 808 State - Oops!.mp3
My Files\Birtney_Speares-Oops...I_did_it_again.mp3
Oops!...I Did It Again.mp3

Fig. 8. An example of misspelling Britney Spears´Oops I did it again. Screen Dump from Napster 05.10. 2001.

5 Discussion

Recently there has been a dispute about what the standards of a P2P file sharing MP3 music may look like. During the last year Napster has become the main tool for downloading MP3 music files, making it a candidate for being the standard within the P2P file sharing community. Its weakness is a lot of unsolved conflicts both towards the record industry and towards other P2P tools. The record industry wants to create a pay-for tool for downloading MP3 files. Napster uses centralized servers for connecting users.

5.1 File Sharing

The basic P2P assumption was about cooperating, equal users sharing network resources. According to a recent report [1] almost 70 % of Gnutella users share no files, and nearly 50% of all responses are returned by the top 1% of sharing hosts showing that free riding is frequent. This may lead to degradation of the system performance. There is a risk for P2P systems being dependent on a few enthusiasts. The increased number of files for CatNap users probably shows this tendency. Such a system may be vulnerable because of lost interest or threats coming from the outside against these main contributors.

5.2 Copyright Protection

Napster will probably meet the demands from RIAA but at the cost of losing the vast majority of the users. As have been shown in Fig. 4 and 5 users leave Napster because there are very few sharing files left. More importantly, those public files left are to a less degree downloaded by users. The difficulties to calibrate the filter made it too efficient. A considerable large proportion of public files did not pass the filter. If the selection of files is reduced, the public or promotional files will be rejected as well. Our guess is that users will use other unfiltered tools rather than a pay-for tool partly because of financial issues and partly because of desires for unlimited search choices.

5.3 Agent Interaction

In the presented work the focus is on investigating the behavior of users downloading MP3-files and to discuss present and future agent interactions. For both the filtering and parasite agents there is a risk of overreacting when trying to optimize MP3 file handling. This is partly due to insufficient calibration and partly to built-in constraints.

Facing the alternatives of joining a pay-for site or participate in collaborative behavior with other Napster users, they may behave more actively in keeping current possibilities for downloading files. Unfortunately this was not possible to prove within Napster because of a too efficient and too unpredictable filter. The Napster filter was tested by changing all filtered filenames by misspelling them within the test set of MP3 files. All filenames passed the filter but within two days they were found and filtered out again.

So, CatNap did not succeed very well in our investigation because the Napster filtering agent detected its strategy early. More fundamentally, there is a twofold risk for CatNap of being too small or too big. Why should a user keep her/his CatNap files if no one else does, or why should Napster keep its server if all users join the CatNap? To avoid disclosure, CatNap should hide from the filtering agent, but that means hiding from everybody else.

The Napster filtering agent aims at satisfying both the record companies and the users sharing MP3 files. At most the filtering agent should strictly filter out all copyrighted files but nothing else. This was not the case in our investigation. It was possible to find a lot of copyright protected music at the same time as all files (both copyright protected and public) at a single file sharer site were filtered out. The filtering agent has a mission impossible, making it a target for both the record companies' pay-for music sites and for other P2P freeloading sites.

5.4 Bandwidth Sharing

When users are shifting from Napster to MusicCity or BearShare, bandwidth sharing will be the next subject of free riding. Both systems exploit faster connections for directing more traffic through these peers. The basic structure of BearShare also causes enlarged overhead because a lot more communication is necessary in a fully distributed system. Both MusicCity and especially BearShare, as shown in Fig. 6, abort more transfers than Napster. A user may waste more and more bandwidth and processor capacity without getting anything back when joining these Napster alternatives. For the Internet as a whole it may end up even worse. Ritter [9] made an arithmetical problem of the Gnutella example presented in Fig. 2. He supposed a user sending an 18 bytes search string, ending up with a total data transmission of over 3 Mb[3]. If a user were allowed to send a request covering a society as big as Napster (we suppose 1 million users) the total data transmission generated would exceed 800 Mb.

5.5 Global Access

File sharing needs global access because a single user does not normally choose who to connect to. The legal actions taken against Napster and in the future probably MusicCity may cause them to close down. There are also possible reactions against BearShare. It is possible to trace IP numbers to a specific BearShare user, since Internet-service providers store the information. Legal notices have already been sent to operators and users are informed they may be threatened legally by the RIAA.

Global access to the core Internet backbone is controlled by Internet backbone providers (IBP). These IBP consist of a few firms[4] mostly located in the US that secure access to the core routing structure, and access to all Internet addresses in the world [4]. Smaller regional Internet service providers are already charged for access to their global infrastructure and core routing services. The problem is also that, as more bandwidth is needed, the strategic importance of those countries that provide it also increases.

5.6 Future

In "the tragedy of the commons" an individual is always supposed to behave selfishly. But individuals within a natural ecosystem have relatives (kin selection within biology) and a possibility to pay and retort a favor from an unrelated neighbor. The success of the ecosystem is dependent on cooperating coalitions. An individual being part of a file-sharing ecosystem will benefit from cooperation if it is possible to identify (real names or pseudonymous) users and/or the cost for doing services is negligible. We found a few such behaviors with users changing file names and joining encrypting programs. In the future, agents may be designed to find other cooperating users or to avoid controlling IBP's, resulting in more or less locally connected groups with their own norms.

A possible future scenario includes but is not restricted to:

- Pay-for music sites are visited by a restricted number of users because people are used to getting downloadable material for free on the Internet.
- The P2P tools providers increasingly get their profit from advertising banners and spywares.

---

[3] An 83 byte data packet sent to 4372 users. The mean number of responses (12%) and the amount of data sent back must also be calculated.

[4] MCI WorldCom, Sprint, GTE, AT&T and Cable & Wireless control between 85% and 95% of the total backbone traffic in the US (Cremer et al 1999)..

- Users have to invest more in bandwidth and technical upgrading.
- New file protection systems like watermarking makes it harder to copy music files but may also reduce the audio quality.

None of the issues above will result in a better-organized file sharing system. RIAA will still have problems getting money out of freeloading users. The new P2P tools will include features not desired by the users. Users have to invest more in technical resources without getting a better audio quality.

6 Conclusions

The two major concerns about P2P are: anonymity and non-censorship of documents. The music industry has highlighted these questions by:

- Forcing Napster to filter out copyright protected MP3 files
- Taking legal actions against local users by monitoring their stored MP3 files

Our investigation shows that when copyright protected files are filtered out, users stop downloading public music as well. When MusicCity and BearShare are replacing Napster, there is an increase in downloading copyright protected files compared to downloading public files. This alteration is contrary to the purpose of introducing a filtering function.

When former Napster users are leaving for other P2P tools, this causes higher bandwidth usage for these users and thus increases the communication needs over the Internet. The Napster alternatives MusicCity and especially BearShare consume more resources both by default and, as shown in our investigation, by having a higher rate of aborted file transfers.

Basically a P2P system consists of a society of equal peers. With central content index, centralized user registration, logon and introduced spywares this is no longer true. There is one group of peers supplying the tools and one group using the tools. Within and between the groups conflicting interests should be expected. The view introduced in this paper treat peers as similar to an (information or biological) ecosystem of agents controlled by Machiavellian beings behind.

The success of a distributed peer-to-peer system is dependent on both cooperating coalitions and an antagonistic arms race. Users may cooperate, i.e. allowing uploading from other users and/or bypass Napster´s filtering function by misspelling or encrypting files. Adar and Huberman [1] have shown the unwillingness for a majority of users to share files and our investigation shows no major tendencies for systematically bypassing Napster´s filtering function. More in general; an individual within the user group will benefit from cooperation if it is possible to identify other users and the cost for doing services towards other users are negligible. In practice it is sufficient to have a fraction of the users cooperating to maintain a well performing system. In Napster there is still some incentive for being cooperating because such a user is not anonymous. In BearShare and MusicCity other users are more anonymous because one MP3 file may be downloaded from multipe sources.

An arms race between antagonistic participants using more and more refined agents is a plausible outcome. Napster´s filtering agent and Catnap, a parasitic agent, are examples of such agents. They fulfill some temporary needs and may probably be replaced by other agents. A possible future agent may be a "bandwidth stealing" agent or a "bandwidth protecting" agent. Users getting used to downloading everything for free has to decide about joining a pay-for service or accepting more power consuming P2P tools with an increased personal effort. Despite legal actions, performance deficits and the strength of commercial forces an uncensored P2P community will probably survive because there are too many new tools developed with innovative new solutions. Instead of "the tragedy of the commons" we are witnessing "the arms race within the commons". Arms race does not need to be a tragedy because these new tools developed or actions taken against too selfish agents may improve the P2P society.

Acknowledgements

I would like to thank Paul Davidsson, Magnus Boman, Ingemar Jönsson and the anonymous reviewers for their comments on various drafts of this work and Martin Hylerstedt for proof reading.

References

1. Adar, A. and Huberman, B.A., Free riding on Gnutella, FirstMonday peer-reviewed journal on the Internet http://firstmonday.org/issues/issue5_10/adar/index.html (2000)
2. Albert, R., Jeong, H., and Barabási, A.-L. Diameter of the World-Wide Web, Nature vol. 401 pp. 130-131 (1999)
3. Carlsson, B. and Davidsson, P., A Biological View of Information Ecosystem, to be presented at IAT 2001, (2001)
4. Foros, Ø., and Kind, H.J., National and Global Regulation of the Market for Internet Connectivityhttp://www.berlecon.de/services/en/iew3/papers/kind.pdf (2001)
5. Hardin, G. The tragedy of the commons, Science vol. 162 pp. 1243-1248 (1968)
6. Hong, T. Performance in Oram A., ed., Peer-to-peer Harnessing the Power of Disruptive Technologies O'Reilly Sepastopol CA (2001)
7. Minar, N., and Hedlund, M., A Network of Peers in Oram A., ed., Peer-to-peer Harnessing the Power of Disruptive Technologies O'Reilly Sepastopol CA (2001)
8. Oram A., ed., Peer-to-peer Harnessing the Power of Disruptive Technologies O'Reilly Sepastopol CA (2001)
9. Ritter, J., Why Gnutella Can't Scale. No, Really. **http://www.darkridge.com/~jpr5/doc/ gnutella.html** 05.31.2001 (2001)
10. Shirky, C., Listening to Napster in Oram A., ed., Peer-to-peer Harnessing the Power of Disruptive Technologies O'Reilly Sepastopol CA (2001)
11. Watts, D.J., and Strogatz, S.H., Collective dynamics of "small world" networks. Nature vol. 393 pp. 440-442 (1998)
12. Williams, G. C., Adaptation and natural selection, Princeton University Press (1966)
13. Wilson, E.O. Sociobiology - The abridged edition. Belknap Press, Cambridge (1980)

Agentspace as a Middleware for Service Integration*

Stanislaw Ambroszkiewicz and Tomasz Nowak

Institute of Science, Polish Academy of Sciences
al. Ordona 21, PL-01-237 Warsaw
and Institute of Informatics, University of Podlasie
al. Sienkiewicza 51, PL-08-110 Siedlce, Poland
{tnowak, sambrosz}@ipipan.waw.pl
http://www.ipipan.waw.pl/mas/

Abstract. Agentspace is an emerging environment resulting from process automation in the Internet and the Web. It is supposed that autonomous software (mobile) agents provide the automation. The agents realize goals delegated to them by their human masters. Interoperability is crucial to assure meaningful interaction, communication and cooperation between heterogeneous agents and heterogeneous services. In order to realize the goals, the agents must create, manage and reconfigure complex workflows. Usually, a workflow integrates a number of heterogeneous services. Our research aims at extracting a minimum that is necessary and sufficient for providing transparency between users and services, i.e. for joining applications as services to agentspace on the one hand and for using and integrating them by heterogeneous agents (on behalf of their users) on the other hand. For this very purpose we introduce a new concept of agent architecture as well as the new agent mobility form called *soul migration*.

1 Introduction

Cyberspace, the emerging world created by the global information infrastructure and facilitated by the Internet and the Web, offers new application scenarios as well as new challenges. One of them is creating new infrastructures to support high-level business-to-business and business-to-consumer activities on the Web, see for example Sun ONE, Microsoft .NET, UDDI, and DAML-S. The infrastructure should provide means for automatic integration of heterogeneous services in the Internet and the Web as well as provide transparency between users and the services. The second challenge is the Semantic Web [8], conceptual structuring of the Web in an explicit machine-readable way. The Semantic Web, to be realized by DAML+OIL, is taken as the basis in the project DAML-S.

It is a common view that autonomous software (mobile) agents can support the process automation in the Internet. Agent is a running program that can migrate from host to host across a heterogeneous network under its own control and interact with other agents and services.

* The work was done within the framework of KBN project No. 7 T11C 040 20. The authors are grateful to Krzysztof Cetnarowicz, Leszek Rozwadowski, Dariusz Mikulowski, Krzysztof Miodek, Jaroslaw Kozlak, and Jacek Gajc for their contribution in uncountable many discussions about the concept of agentspace.

A. Omicini, P. Petta, and R. Tolksdorf (Eds.): ESAW 2001, LNAI 2203, pp. 134–159, 2001.

Since the software agents are supposed to "live" in the cyberspace, they must be intelligent, that is, they must efficiently realize the goals delegated to them by their human masters. Hence, along the development of cyberspace the new world (called agentspace), inhabited by the software agents, is being created.

Human users are situated at the border of the agentspace and can influence it only by their agents by delegating to them complex and time consuming tasks to perform. Since the Internet and the Web are open distributed and heterogeneous environments, agents can be created by different users whereas services by different providers according to different architectures. Interoperability is crucial to assure meaningful interaction, communication and cooperation between heterogeneous agents and services. We can distinguish two kinds of interoperability: Interaction interoperability and semantic interoperability. Interaction interoperability provides common communication infrastructure for message exchanging whereas semantic interoperability provides the message understanding.

The semantic interoperability concerning the meaning of resources on the Web is a subject of current research, see DAML[12] + OIL[20] as the most prominent example.

In order to use services established by different users working in heterogeneous domains, agents must be capable of acquiring knowledge about how to use those services and for what purposes. There must be a common language for expressing tasks by the users, delegating these tasks to agents, as well as for describing and integrating services, and for communication between agents and services. There are several efforts for creating such language: LARKS [22], ATLAS [7], CCL [25], WSDL [24], FIPA ACL [13], and quite recently DAML-S [18]. The semantic interoperability of DAML-S is based on the explicit meaning to be provided by ontologies constructed in the language DAML+OIL. We also propose a language, called Entish, however the semantic interoperability of our language is different. We follow the idea of Wittgenstein [27] that the meaning of language is in its use. We construct means for providing such semantic interoperability. Our approach seems to be extremely simple and efficient. The key idea is that an agent needs not "know" explicit formal meaning of resources and functions performed by services. The agent must only know how to use the resources and services that are necessary to realize the task delegated by user.

As to the communication infrastructure, there is no need to force one transportation platform (i.e. one message format and one message delivery way) as the standard. It seems that rather message language and its meaning is crucial here, not message wrapping. It is relatively easy to provide a transformation service between two platforms for translating message format of one platform to the message format of the other.

Mobile agent platform (MAP, for short) gives also a communication infrastructure as well as "migration service" for the agents. One may ask if agent mobility is essential for creating agentspace, see for example JADE [14] framework where mobility is not provided. In our approach, agent mobility may be seen as a means for learning between heterogeneous environments.

Our project aims at creating an absolute minimum necessary for joining heterogeneous applications as services on the one hand and for integrating and using them by heterogeneous agents (on behalf of their users) on the other hand. As this minimum we propose the language Entish (a shorthand for e-language), and its intended semantics.

Agent mental attributes (i.e., goals, intentions, commitments, knowledge) constitute the core of Entish. This gives rise to a new concept of agent architecture. The idea of the architecture is that agent is created and dedicated to a particular task. The agent is responsible only for constructing and executing a workflow that integrates services needed for the task realization. The architecture is distributed because planning and reasoning capabilities are situated outside the agent as well as the capability to perform actions. The only action our agent can perform is communication with services and possibly migration to another host over the Internet. All the essential data of agent functioning are stored in its mental attributes. These attributes form agent soul that is separated from agent mind responsible for decision making, and from agent body responsible for action execution and environment perception. One of the main advantage of the new architecture is that the essential data of agent functioning are stored in its soul, so that the agent process may be closed at any time. Since the soul is independent from mind and body, it can be moved to another place and given another mind and body (i.e., agent process can be fully reconstructed) in another place. This gives rise to introduce *soul migration* as a new agent mobility form. The concept of soul solves several hard problems of mobile agent technology such that the agent persistence, and the security of hosts open for strange agents. The agent persistence follows from the concept of soul where all essential data of agent functioning are stored. Since during agent migration (actually soul migration) only plain data (not binary code to be executed) are send to a remote host, the problem of host security is also solved.

Probably the most related work is PageSpace [10,11]. PageSpace is a reference architecture for multiagent applications built on top of the Web. Agentspace can also be built on HTTP transport as it is in the case of our ongoing implementation of the communication platform called Hermes. In our framework we have only one kind of agents that correspond more or less to the application agents in PageSpace. The other kinds of PageSpace agents correspond to services in our framework, i.e., the PageSpace agents perform similar functions as our services do. However, the main difference is the language for communication and coordination. In PageSpace the language is based on Linda, whereas in our approach it is Entish that is a fully declarative language. Although the goal of PageSpace is the same as the goal of agentspace, i.e., service integration by autonomous agents, the design principles and chosen architectures are completely different.

The main achievement of our project is a generic architecture of agentspace and its implementations. The idea of agentspace consists in constructing middleware that provides transparency between heterogeneous agents and heterogeneous services. We define agentspace as an implementation of the language Entish and its intended semantics on a communication platform. As a consequence of agentspace design principles, we propose also a new concept of agent architecture.

The paper is structured as follows. In Section 2, a motivation of our approach is given. In Section 3, the layered architecture of agentspace is presented. Section 4 is devoted to presentation of agentspace as a middleware. In Section 5, language layer of agentspace architecture is presented. Section 6 gives an overview of agent/service layer. Section 7 is devoted to a short presentation of interaction layer. The syntax and intended semantics of Entish is presented in the Appendix (Section 9).

2 Motivations

Perhaps in the near future the development of agentspace reaches the critical mass, i.e. the number of connected applications will be so large that a user, facing hard, complex and time consuming task to perform, first of all will check if this task could be realized in the agentspace. So that she/he will formulate the task in Entish, delegate this task (together with a timeout) to an agent, send the agent to the agentspace and wait for a result. On the other hand if a programmer has implemented an application that performs complex and time consuming routine, and the interface needed to connect it to the agentspace infrastructure is easy to implement, why not to do so?

Suppose that a user faces the following problem. He or she has a file in latex format and wants to transform it onto HTML format. However, the user has not installed 'LATEXto HTML' converter on his workstation. Since the job is to be performed only once, the user does not want to take all this burden of purchasing the system or looking over the Web to download a free copy of it, and then installing it. Supposing that there is a service in agentspace that performs that job, the user formulates the task, describing the job, by using SecretaryService - his interface to the agentspace. The task description is expressed in our language Entish that is supposed to be the common language of the agentspace.

Let us present a rough scenario that may follow from that situation. Suppose, the user delegates the task (along with a timeout) to an agent, and the agent is sent into agentspace. The agent is autonomous and its only goal is to realize the task. The task is expressed in Entish as formula, say f, describing a class of situation that realize the task. Suppose that the agent is equipped with the initial resources described by the Entish formula f^o. First of all, the agent gets in touch with an information service (called InfoService) and sends the following message: *"my intention is f"*. Suppose that the InfoService replies with the following: *"service s^* can realize your intention"*. So that the agent sends again the message *"my intention is f"*, however this time to the service s^*. If the service s^* can and is ready to do so, it sends back the following message to the agent: *"s^* commits to you to realize f, if f^* is realized"*.
Let us note that here the formula f refers to the output of the service s^*, whereas f^* to the input. Having the commitment, the next intention of the agent is to realize f^*, i.e., the input of s^*. So that the agent sends the message *"my intention is f^*"* to any known InfoService. It the agent gets no reply from the InfoServices, it migrates to any randomly chosen remote place and tries again to communicate with new InfoServices. Suppose that finally, the agent got back the following: *"service s^{**} can realize your intention f^*"*. The agent sends to the service s^{**} the following message: *"my intention is f^*"*. The service replies with the following: *"s^{**} commits to you to realize f^*, if f^{**} is realized"*. Here again the formula f^* refers to the output of service s^{**}, whereas f^{**} to the input. Now, suppose that f^{**} follows from f^o. So that having these two commitments and knowing that f^{**} follows from f^o, the agent has already arranged a simple workflow needed to realize the task delegated by the user. If the workflow is executed and performed successfully, the agent notifies the user about that and can finish its activity. However, before the agent does so it is obliged to report its experience (i.e., the operation it has performed in order to realize the delegated task) to an InfoService. In this way the system has ability to learn, so that the next agents having tasks of the

same type will be informed how to realize them immediately. If the workflow fails, the agent makes its best to recover it and tries again unless the timeout is over. In the case the timeout is over, the agent notifies the user about failure and also finishes its activity.

Although our example may be seen as a bit abstract one, it gives an idea of the simple universal conversation protocol (see Appendix for details) describing how distributed heterogeneous services can be integrated (composed) into a workflow needed for task realization. Another example from e-commerce domain can be given, however the limit of space does not allow to do so.

The conclusion from the example presented above is that formal ontologies are not necessary for providing semantic interoperability in the domain of service integration. Our approach to semantic interoperability may be summarized as follows. We need semantic interoperability only for the following purpose:

- for automatic integration of heterogeneous services as well as
- for providing transparency between users and services.

It is clear that such semantic interoperability may be realized in different ways. One natural way to do so is based on formal ontologies. However, it is becoming clear that this way is hard to implement and perhaps even much harder to apply in large scale. Our idea is that the purpose can be realized in an extremely simple way. What is needed is a common open declarative language that can be used and developed in a distributed way by users, agents and service providers. We propose Entish as the language. Agentspace, i.e., an implementation of Entish on a communication platform, provides minimum infrastructure necessary to realize the purpose.

3 Agentspace Architecture

The idea of agentspace consists in construction of open distributed infrastructure that would allow to join heterogeneous applications as services on the one hand and to use them by heterogeneous agents on the other hand. A user, delegating a task to an agent, needs not to know the locations of services and resources necessary for task realization. The user expresses only the task in our high level common language Entish. The agent migrates across the agentspace looking for information, services and resources needed to realize the delegated task.

Since agentspace is an implementation of the language Entish and its intended semantics on a communication platform, the layered architecture, shown in Fig. 1 seems to be natural and generic. The architecture consists of three layers: interaction layer, agent/service layer, and language layer. The interaction layer specifies infrastructure that provides basic functionality for agents and services like agent moving from one place to another, communication between agents and services as well as agent perception of the environment. This layer is implemented by a communication platform. In our case it is done by Pegaz and Hermes. However, it may be any communication platform, like JADE [14], or a new one built on, for example, on the top of CORBA, RMI-IIOP.

The second layer, i.e., agent/service layer, specifies some aspects of agent and service architecture that allow them to evaluate formulas (called situations) expressed in the language Entish as well as determining new situations resulting from performing

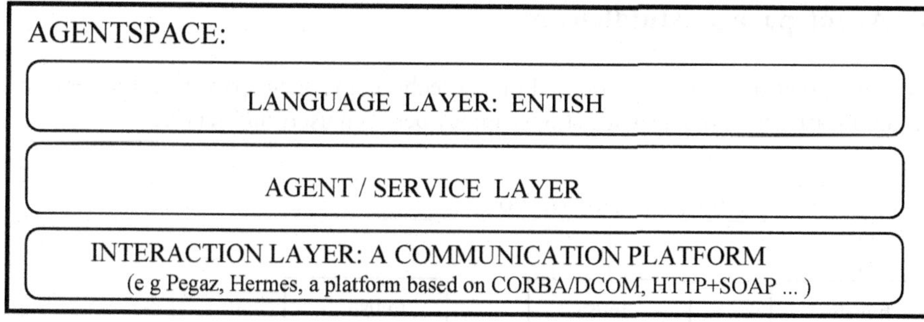

Fig. 1. The Layered Architecture of Agentspace.

elementary actions. The agents are equipped with mental attitudes: knowledge, goals, intentions and commitments represented as Entish formulas. These attitudes serve as data and control parameters of agent behavior. Agents execute actions (migration and message exchange) in the interaction layer, whereas the message contents is expressed in Entish. The agent/service layer implements the intended semantics of Entish.

The language layer consists of Entish - a simple version of the language of first order logic along with a specification how to "implement" it for open and distributed use. The implementation follows the idea of so called "webizing language" see T. Berners-Lee [8]. The language describes the "world" (i.e., agentspace) to be created on the basis of infrastructure provided by the previous layers. However, this description is purely declarative. Actions are not used in Entish; the formulas describe only the results of performing actions. So that no dynamics and no causal relations can be expressed here. The language is sufficient to express desired situations (tasks) by the users as well as by agents and services, however it can not explicitly express any idea about how to achieve them. This may be done by implementing distributed information services (called InfoServices) where agent can get to know how to realize the delegated task, or get a hint. Usually, as the reply to its query (expressed also in Entish) agent gets a sequence of intermediate situations to follow.

The language is implemented in the second layer by DictionaryServices containing the syntax and new concept definitions. There are also three additional types of services, namely SecretaryService (a GUI), MailService (for asynchronous communication), and BodyService for soul migration. Let us note that all those services are not system services. They can be implemented and developed independently by different users. It is important that only "operation type" of any of these services is specified in Entish. Operation type is a description of the function performed by a particular service. A service implementation must only satisfy specification of the operation type. The details of language layer as well as Entish syntax are in the Appendix.

4 Agentspace as Middleware

From the point of view of distributed object technology, agentspace is a middleware, i.e. an infrastructure that provides transparency between users and services.

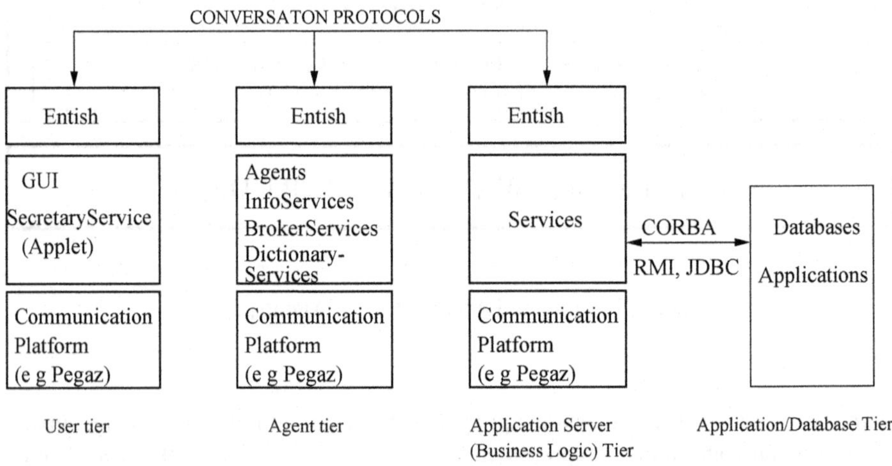

Fig. 2. The Multi-tier Architecture of Agentspace.

Applications built using distributed objects such as CORBA naturally lend themselves to a multitiered architecture. A typical three-tiered application has a user interface tier, a computation (or business logic) tier, and a database access tier. All interaction between the tiers occurs via the interfaces that all CORBA objects must publish. The user-interface tier invokes methods on the business logic tier and thus acts as a client of the business logic servers. The business logic layer is made up of business objects - CORBA objects that perform logical business functions such as inventory control, budget, sales orders, and billing. These objects invoke methods on Data Store tier objects that encapsulate database routines.

Another distributed application concept is proposed by SUN. It is The JavaTM 2 Platform, Enterprise Edition. The architecture of this platform is similar to the multi-tier architecture presented above where CORBA is substituted by RMI-IIOP.

It seems that the agentspace architecture may be seen as a proposal of an alternative multi-tier architecture for distributed applications. We introduce two new components. The first one is a universal communication language that serves for much more than a language for simple method call of remote heterogeneous objects, like IDL of CORBA. Entish can be used to specify input and output type of a service, function performed by a service as well as timeouts for service performance. The second new component is the additional tier between user tier and business logic tier. It is the agent tier that serves for constructing new (if needed) composed services on the basis of the existing ones as well as for finding out resources and new services for realizing tasks delegated to agents by the users. This agent tier allows to realize the tasks in an open and dynamic environment like the Internet and the Web. If the environment is closed like a classic enterprise system, then the three-tier middleware architecture (for example, J2EE) is sufficient. In such closed environment, object locations and their method specifications are known a priori, so that an additional tier is not necessary here. However, in an open, dynamic and heterogeneous environment where large number of programmers modify existing objects and join new ones, an additional coordinating tier is necessary as well as common, platform-independent communication language capable to express the essence of data processing.

5 Language Layer: Entish

Our goal is to specify and implement minimum infrastructure necessary to create agentspace. It is important to note that this minimum is not a specification of the architecture presented in Fig. 1 and Fig. 2. The minimum consists only of the language and its intended semantics. The minimum is also independent of implementation of interaction and agent/service layers. However, the interaction and agent/service layers can not be arbitrary, they must satisfy several general conditions imposed by the semantics.

The language layer consists of the language, called Entish, of first order logic with types, and a specification how to implement it in open distributed environment using space of unique names URI (Universal Resource Identifiers), as suggested by T. Berners-Lee [8]. There is only one deviation from the classic logic; we treat formulas as objects in our language. It is a convenient way to express agent mental attitudes, like intentions, goals, knowledge, and commitments.

Without going into details, Entish consists of types of primitive objects that include: places, agents, resources, services, agent/service mental attributes, and Time. There are also two primitive relations: is_in and leq, and several primitive functions, like $gmt()$. The first relation expresses that something "is in" something else, for example, an agent "is in" a place. The second relation is needed to express timeouts, for example, that a date (object of type Time) is less or equal (leq for short) than the current GMT time delivered by function $gmt()$. The rest part of Entish consists of standard first order logical operators, for term and formula constructions.

In order to realize the soul migration, the special type of resources is necessary. It is called Soul. A resource of that type, a soul, stores all essential agent's data: its name, goals, intentions, commitments, knowledge, history. Since soul is a resource, it can be sent form one place to another during a communication session. Once a soul is given a body and becomes a running process, it is identified with the agent whose name is stored in the soul.

The main purpose of Entish is to provide simple common conversation language for users, agents, and services. The conversation consists in passing a message, i.e., resource "info" of special type called Info. Intuitively, this info carries a fact, i.e., an evaluated Entish formula. Using the mental attitudes, a conversation protocol can be constructed in the following way. Agent, say *ag1*, always sends to a service a message containing a formula of the form: *intentions(ag1) implies* ϕ, where ϕ is an Entish formula. The intended meaning is that ϕ follows from the current intention of agent *ag1*. If

the service (say *ser*) agrees to realize agent's intention, it sends back an info containing its commitment expressed as the following Entish formula:
(ψ implies form_in(commitments(ser)) and
(form_out(commitments(ser)) implies ϕ)
Without going into the notation details, it means that the service *ser* commits to realize formula ϕ if only the formula ψ becomes true. However, if the service *ser* is an InfoService, it sends back an info containing formula of the following form:
form_out(operation_type(cos)) implies intentions(ag1), where *cos* is the pointer (universal address, i.e., a URI) to a service or to a (partial) plan that may realize agent *ag1*'s intention. The conversation protocol is completed by four additional control messages needed for workflow management; see the Appendix for details.

5.1 Formal Model of Agentspace and Semantics of Entish

Formal model of agentspace is based our previous paper [3]. Agentspace is an implementation of Entish and its intended semantics on a communication platform. Communication platform can not be arbitrary; it should provide places for services and for agents that can move from place to place. The agents and services should be able communicate, i.e., to exchange data and resources. It seems that this constitutes the minimum necessary to implement Entish.

Since Entish formulas do not describe actions, the set of actions that can be executed on communication platform is not determined. In Pegaz, there are the following basic types of primitive actions: *create agent*, *communicate*, *migrate*, *send (resource)*, *terminate*. However in Hermes, migration is not a primitive action because Hermes is not a MAP, so that agent migration (along with its code) is not possible there. In Hermes, the migration is realized by the actions *send, terminate* and BodyService, that is, agent's soul is sent as a resource to a BodyService located on the new place where the agent wants to migrate. Then, the BodyService creates the agent having this soul whereas the original agent terminates its process on the old place.

Let us notice that only the action *terminate* can be performed individually by an agent. The rest of the actions are joint ones. Joint action a (for example a communication action) can be executed if, for all the needed agents/services/resources/places "can" and "do want" to participate in the execution of this action. If one of them cannot or doesn't want to participate in the execution, then the attempt to execute the action a fails.

The crucial notion, needed to define a model of agentspace, is the notion of "event". Any event corresponds to an action execution, so that it "describes" an occurrence of a local interaction of entities participating in the execution. The term "local" is of great importance here. Usually, local interaction concerns only few entities (up to five) so that the event associated with this local interaction describes only the involved agents, services, resources, and places. The events form a (partial order) structure, that expresses their causal relations.

Having primitive concepts and setting initial conditions for agents, places, and distribution of primitive resources and services, the set of all events that can occur in the environment can be defined. It is important to notice that due to the local interaction, it is easy to extend the set of events if new agents, places, resources, and services are added.

Hence, the representation is appropriate for open environments. Event structures have been successfully applied in the theory of distributed systems [26] and several temporal logics have adopted them as frames [17].

The event structure is a formal model of agentspace. Agent mental attitudes (knowledge, goals, intentions, and commitments) are represented by sets of runs of the event structure. This allows to specify, and verify agent/service behavior in terms of these mental attitudes; see [3] for details. Formal semantics of Entish is defined as a specification of agent/service behavior.

In order to have a complete picture of what is going on, let us consider Fig. 3 showing the relation between world, model, and language. The language describes the model whereas the model is an abstraction of the world. In our case the world is Cyberspace. The model is the event structure presented above; it is an abstract view of the Cyberspace via a communication platform (Pegaz). The language is Entish that describes the static properties of the model.

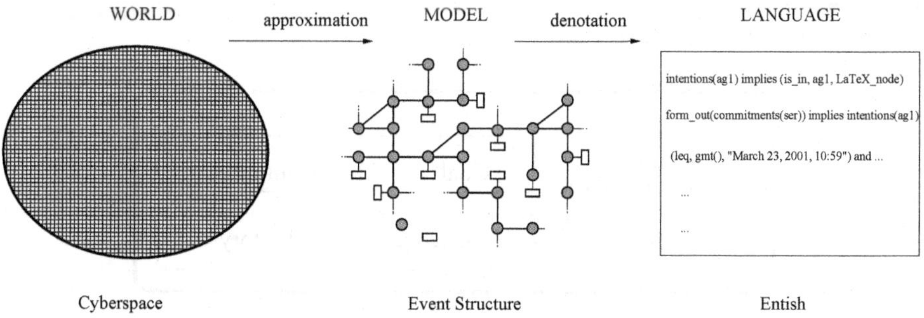

Fig. 3. The Relation between World, Model, and Language.

Let us conclude this section with the following important remark. Entish primitives are only places, agents, resources, services, agent/service mental attributes, Time and relations is_in and leq. It seems that these primitives are general enough to capture the essence of agent paradigm as well as data processing. Since Entish does not contain actions, the set of actions in the model can be arbitrary, i.e., not necessarily corresponding to the Cyberspace. So that the class of models the Entish can describe is larger than the class of Cyberspace models. Hence, it seems that Entish may be regarded as a universal agent language.

Formal intended semantics will be presented in a separate technical paper. Informal semantics of agent/service behavior is given in the next section whereas the syntax of Entish is presented in the Appendix.

6 Agent/Service Layer

This layer implements Entish and its intended semantics (i.e., the behavior of agents and services) on a communication platform. The basic components of the layer are agents and services. Besides ordinary services which are applications joined to agentspace, we can distinguish auxiliary services that are devoted to help the agents to realize their goals. They are not system services so that their implementation is delegated to users. We propose and specify the following auxiliary services: DictionaryService, InfoService, BodyService, SecretaryService, MailService. The details on these services are in the Appendix. The idea of BrokerService is presented in Section 6.3.

6.1 Agent Architecture

Agent is a running process, created by a user or a service, that can migrate across an agentspace and between agentspaces built on different communication platforms. User or service delegates to agent a goal to realize. The goal is realized by construction and execution of a workflow that usually integrates a number of heterogeneous services. In order to provide such functionality we propose the following agent architecture.

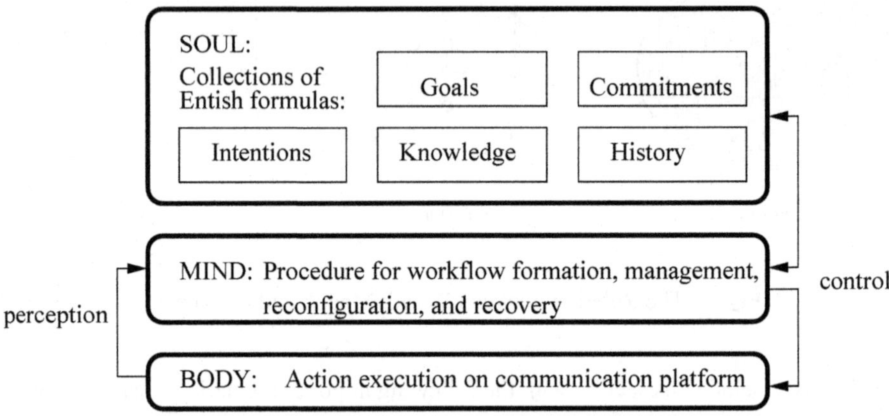

Fig. 4. The Layered Agent Architecture.

Agent architecture is composed of the following three layers: mental attitudes (soul), decision mechanism (mind), execution mechanism (body). The soul is expressed in Entish and consists of knowledge (a collection of facts), goals, intentions, history and commitments. Since the contents of soul is expressed in Entish, it (the contents) is independent from communication platform as well as from agent implementation. Agent soul can migrate alone without mind and body. This gives rise to introduce new form of agent mobility called *soul migration.* The idea is that a running agent process stores all its data and control parameters in its soul. The process may be closed at any time and then fully reconstructed at any new place on any communication platform. At the new place, agent soul may be given any mind and body and then the completed agent can continue its process. The new agent migration form is based on one common format of agent data (soul). Usually, agent data structure is arbitrary and depends on agent implementation, so that migration of agent data makes sense only if the same agent code is available at the new place. In our approach agent data has one universal structure so that it is compatible with any agent code that implements the common format of soul. Although the soul migration is the basic migration form in our architecture, within the same MAP agent can migrate as one integral entity, i.e., consisting of soul, mind, and body.

In order to realize the soul migration, the special kind of data structure is necessary. It is called Soul. An instance of that kind, a soul, stores all essential agent's data: its name, goals, intentions, commitments, knowledge, history. Since soul is a data, it can be sent form one place to another during a communication session. Once a soul is given a mind and a body and becomes a running process, it is identified with the agent whose name is stored in the soul. The soul migration is realized by special service called BodyService.

The mind controls both the soul and body. Agent body executes actions (directed by the mind) in the interaction layer. It is left open to a programmer how to design and implement his own agent mind and body. However, to assure interoperability, the programmer must submit to the one common soul format. The specification of the common soul format is given in the Appendix.

Although the soul format is based on BDI agent architecture [9,21], from the point of view of mobile agent technology, the proposed soul format may be insufficient, that is, some important aspects (data) of agent process can not be stored in the format. The soul format was design for agents that are supposed to integrate heterogeneous services. So that from this point of view it works well. Perhaps the proposed format is not universal so that in order to apply our agent architecture in other domains, the soul format should be extended.

New agent is created if a new task is delegated by a user. First of all, agent soul is constructed by writing goal, knowledge, and commitment formulas. The agent goal expresses the task. The knowledge is a collection of facts that give the agent initial knowledge about, for example, resources provided by the user for task realization. The commitment means that the agent is bound to realize the task for the user. Now, only agent mind and body is needed to complete the agent and send it as an autonomous entity to the agentspace. In order to do so the soul is send to a BodyService which activates an agent process based on the soul. In this way the agent begins its esoteric life.

A consequence of our agent architecture is that agent is purely an information agent. Although we allow agent to carry resources, it is not recommended especially if the agent is supposed to migrate (it is autonomous, so it can do so) to unknown platforms. Its capability to perform actions is restricted mainly to migration and communication however, there is a possibility of having more actions implemented as routines in agent body. Its ability to reason is minimal, almost all reasoning and planning job is delegated to the InfoServices.

So, the question is what our agent does? The main and only job of the agent is to construct workflow that would perform successfully the task delegated to that agent.

This includes: (1) construction of operation (a sequence of sub tasks) of the workflow; (2) arrangement of services needed to perform the sub tasks; (3) control, rearrangement and recovery of the workflow in the case of failure; (4) and finally notification (either positive or negative) of the task realization to be sent to user.

Agent decision mechanism can work according to the following algorithm:

1. update knowledge on the basis of perception;
   check all your timeouts;
2. check if your current intention and/or goal is realized;
   if the goal is realized or the timeout to realize it is over, go to (7);
   if the current workflow execution fails, go to (6);
   if the current intention is realized go to (6);
3. check if there is a routine (a primitive action) in your body that can realize your current intention;
   if there is one, execute it and go to (1);
4. ask a service how to realize the intention;
   if you get back a partial plan, go to (6);
   if you get back a commitment, take the precondition of the commitment as the current intention;
   if the precondition is true, go to (5), otherwise go to (1);
   if all known InfoServices can not help you, migrate randomly to another remote place (to look for another InfoServices) and go to (1);
5. if the workflow is completed, execute it and go to (1);
6. planning and determining (next) current intention;
   go to (1);
7. notify (positively or negatively) the user;
   give a report to an InfoService;
   terminate the process.

It seems that our agent architecture can not be classified according to the standard taxonomies, see for example [19,23]. Perhaps it is specific for application domain, that is, for Internet and Cyberspace. Our architecture may be seen as distributed, i.e. InfoServices may be viewed as part of agent architecture where the main part of agent planning and learning is performed. The learning consists on storing and processing agent experience. Agent life is short, i.e., it "dies" after realizing its task (or if the timeout is over) and reporting the way the task was achieved to an InfoService. On the other hand, it is easy to create a new agent if there is a task to be realized. The new agent can use the experience (of the past agents) stored and processed in the InfoServices.

6.2 Ordinary Services

It is of great importance in our framework that *any application* can be joined to agentspace as an ordinary service. In order to do so, the function performed by this application must be expressed in Entish, and the service must have an interface between the application and agentspace. The interface is responsible for communication as well as for transporting input data to the application, and output data from the application. The interface implements simple common conversation protocol (for details see Appendix, section on Entish) on a communication platform. The implementation is delegated to service provider who wants to join his application to agentspace. The problem is when the function performed by the application can not be expressed in the current Entish. However, Entish is an open language, and the user can create his own name for the function and publish it in a DictionaryService. In this way the function name becomes a new function symbol in Entish. Of course, it is better for the new application if it is a new and perhaps better implementation of the already existing function defined in Entish, or it is an implementation of a function being a composition of the existing functions. The provider is obliged also to create (according to the URI protocol, see Appendix, Section 9.5) the complete name (URI) for his service. Then, the service provider should advertise its application by sending info to InfoServices about the operation type performed by the service. Finally, the new service is ready to perform tasks, so that it waits for clients.

6.3 BrokerServices

Our idea of BrokerService is similar to the broker agent, see [16] for overview. Broker agent is considered there as a special type of middle-agent that provides mediation service between service provider agents and service requester agents. A broker agent provides all communication, negotiations, executions, and controls of transactions with service provider, and then returns final result to service requester.

Another types of middle-agent are mediator agent and matchmaker agent. In general, a mediator focuses on the provision of services devoted to semantic information brokering whereas a matchmaker simply returns a ranked list of relevant service providers to the requesting agent.

The principles of our architecture are different. Our agents can be viewed as service requester according to [15]. They are given tasks to perform. The tasks may be complex and need a construction and management of workflow that may engage a quite large number of services. So that usually our agent requests more than one single service in order to realize its task. Mediator and matchmaker of [15] are implemented as InfoServices in our architecture.

Broker agent of [15] is implemented as BrokerServices in our architecture. The idea of BrokerService is the following. The service gathers information about task performance demand in the agentspace. If tasks of some type are frequently performed by agents, the service constructs and manages permanent workflow that performs that type of tasks. All this information, i.e. about the task type and how to construct the workflow, is given to the BrokerService by InfoServices. So that the role of InfoServices is extremely important in our architecture.

Once the workflow is established, the BrokerService is ready to realize orders. The service advertises the type of tasks it realizes in InfoServices, and waits for clients. So that, now agents need not to construct and manage workflows by themselves. An agent goes directly to the appropriate BrokerService that is ready to perform the task for that agent. The service takes also care of workflow management, so that the agent is "happy" to delegate its task to the service.

If the number of agents requesting this type of service significantly decreases, the BrokerService closes the workflow, and starts again to listen what type of service is needed in agentspace now.

It is clear that in order to manage the workflow efficiently, the BrokerService must perform all functions typical for organization. The problem is that even a simple organization has sophisticated structure, in which we can distinguish: the goal of the organization, business process realizing that goal, internal information flow, and organizational dependencies concerning the decision power distribution.

The architecture of BrokerService is based on our concept of agent virtual organization [4], and agent virtual enterprise [5,6].

7 Interaction Layer

Interaction layer is implemented as a communication platform. As it was already mentioned, the communication platform need not be a MAP. So that the question is: What are the minimum requirements for the platform? It seems that URIs (universal addresses) for resources and for processes (services and agents) running on places, and a communication between these processes as well as agent perception functions are the sufficient requirements. By agent perception available at a place we mean the following: the name of place, GMT, list of services on the place, vicinity of place (list of places associated to that place), and may be some others. The perception functions are necessary for agent to evaluate Entish formulas.

Hence, communication platform must provide places, possibility to run service processes and agent processes on that places, a communication between them, and agent perception. Since the requirements for communication platform are simple, it can be built, for example, on HTTP+SOAP transport.

Summing up, the interoperability necessary for agent integration of heterogeneous services into workflows and their management can be provided by the following:

- Entish and its intended semantics as well as the agent architecture that results from Entish,
- the minimum requirements for communication platform presented above.

These two items constitute the core of our work. Different implementations of the agent/service layer on communication platforms results in different agentspaces. Since Entish and specification how to implement it on agent / service layer are supposed to be fixed and common, the problem of interoperability between different agentspaces is reduced to the interaction interoperability, i.e., interoperability between communication platforms. This can be realizes in a simple way by implementing translation between the transport protocols used in the communication platforms.

8 Conclusions

The paper presents our work in progress. The limit of space does not allow to present details. The first version of Entish syntax and intended semantics is completed and presented in the Appendix. A prototype of agentspace based on Pegaz is already implemented as well as a prototype of our agent architecture. Implementation of Hermes, i.e., agentspace based on HTTP transport, will be completed shortly. Now, we are developing (by implementing services) and testing our small agentspace in the frame of Pegaz Ring that consists of several research groups.

References

1. S. Ambroszkiewicz. Interoperability in Agentspace: proposal of agent interface to environment. In Proc. of Workshop on Semantic Web: Models, Architectures and Management `http://www.ics.forth.gr/proj/isst/SemWeb/` at Fourth European Conference on Research and Advanced Technology for Digital Libraries, 18 – 21, September 2000.
2. S. Ambroszkiewicz. Towards Software Agent Interoperability. In Proc. 10th Conf. on Information Modelling and Knowledge Bases, Saariselka, Finland, May 8-11, 2000. Extended version will appear in Kangassalo H., Jaakkola H., (Eds.) Information Modelling and Knowledge Bases XII, IOS Press, Amsterdam, 2001.
3. S. Ambroszkiewicz, W. Penczek, and T. Nowak. Towards Formal Specification and Verification in Cyberspace. Presented at Goddard Workshop on Formal Approaches to Agent-Based Systems, 5 - 7 April 2000, NASA Goddard Space Flight Center, Greenbelt, Maryland, USA. To appear in Springer LNCS.
4. S. Ambroszkiewicz, O. Matyja, and W. Penczek. "Team Formation by Self-Interested Mobile Agents." In Proc. 4-th Australian DAI-Workshop, Brisbane, Australia, July 13, 1998. Published in Springer LNAI 1544.
5. S. Ambroszkiewicz. Agent Virtual Organizations within the Framework of Network Computing: a case study, In Proc. CEEMAS'99, The First International Workshop of Central and Eastern Europe on Multi-agent Systems, 1st-4th June 1999, St. Petersburg, Russia.
6. S. Ambroszkiewicz, K. Cetnarowicz, J. Kozlak, and W. Penczek. Modeling Agent Organizations. In Proc. of Special Session on Agent-Based Simulation, Planning and Control of the 16th IMACS World Congress 2000 on Scientific Computation, Applied Mathematics and Simulation EPFL, Lausanne, Switzerland, 21-25 August 2000
7. ATLAS - `http://www.cs.cmu.edu/~softagents/atlas/`
8. T. Berners-Lee - `http://www.w3.org/DesignIssues/Webize.html` -and- `/DesignIssues/Logic.html`
9. M. E. Bratman. Intentions, Plans, and Practical Reason. Harvard University Press, 1987.
10. P. Ciancarini and R. Tolksdorf and F. Vitali. The World Wide Web as a Place for Agents. In M. Woolridge and M. Veloso (Eds.) *Artificial Intelligence Today. Recent Trends and Developments,* Springer LNAI 1600, 175-194, 1999.
11. P. Ciancarini and R. Tolksdorf and F. Vitali and D. Rossi and A. Knoche. Coordinating Multiagent Applications on the WWW: a Reference Architecture, *IEEE Trans on Sw Eng,* Vol 24, No 5, 362-375, 1998.
12. DARPA Agent Markup Language DAML `http://www.daml.org/`
13. FIPA - The Foundation for Intelligent Physical Agents, `http://fipa.org/`
14. JADE - Java Agent DEvelopment Framework `http://sharon.cselt.it/projects/jade/`
15. Klusch, M.: Information Agent Technology for the Internet: A Survey. Journal on Data and Knowledge Engineering, Special Issue on Intelligent Information Integration, D. Fensel (Ed.), Vol. 36(3), Elsevier Science, 2001.
16. Klusch, M., Sycara, K. Brokering and Matchmaking for Coordination of Agent Societies: A Survey In: Coordination of Internet Agents, A. Omicini et al. (eds.), Springer, 2001.

17. K. Lodaya, R. Ramanujam, P.S. Thiagarajan. "Temporal logic for communicating sequential agents: I", Int. J. Found. Comp. Sci., vol. 3(2), 1992, pp. 117–159.
18. McIlraith, S., Son, T. and Zeng, H. "Mobilizing the Web with DAML-Enabled Web Services", `http://www.ksl.stanford.edu/projects/DAML/` and www site of DAML-S: `http://www.daml.org/services`
19. J.P.Mueller. The Right Agent (Architecture) to Do the Right Thing. In J.P. Mueller, M.P. Singh, and A.S. Rao (Eds.) *Inteligent Agents V, Proc. of ATAL'98,* Springer LNAI 1555, pp. 211-225, 1999.
20. OIL, Ontology Interchange Language, `http://www.ontoknowledge.org/oil/`
21. A. S. Rao and M. P. Georgeff. Modelling rational agents within a BDI–architecture. In Proc. KR'91, pp. 473-484, Cambridge, Mass., 1991, Morgan Kaufmann.
22. Sycara, K.; Widoff, S.; Klusch, M.; Lu, J.: LARKS: Dynamic Matchmaking Among Heterogeneous Software Agents in Cyberspace. Journal on Autonomous Agents and Multi-Agent Systems, Kluwer Academic, March, 2001.
23. W. Truszkowski and J. Karlin. A Cybernetic Approach to the Modeling of Agent Communities. In Proc. of 4th International workshop CIA 2000, Boston, MA, USA, July 2000. LNAI 1860, pp. 166-178.
24. Web Services Description Language (WSDL) 1.1 - `http://www.w3.org/TR/2001/NOTE-wsdl-20010315`
25. S. Willmott, M. Calisti, B. Faltings, S. Macho-Gonzalez, O. Belakhdar, M. Torrens. "CCL: Expressions of Choice in Agent Communication". In Proc. of ICMAS-2000.
26. Winskel, G., An Introduction to Event Structures, LNCS 354, Springer - Verlag, pp. 364–397, 1989.
27. L. Wittgenstein. *Philosophical Investigations.* Basil Blackwell, pp. 20–21, 1958.

9 Appendix: The Syntax and Intended Semantics of Entish

The design of Entish aims at realizing the idea of simple protocols (presented in the previous sections) for constructing, executing and managing workflows by agents in order to realize the tasks delegated to them by users.

The language Entish is composed of three parts: Upper Entish, Proper Entish, and Communication Entish; for short uEntish, pEntish, and cEntish.

uEntish is a simple version the language of first order logic with types without quantification. pEntish is an instance of uEntish where specific types, relation and function names, as well as atomic formulas are introduced. cEntish is a simple communication language built on the top of pEntish.

9.1 Upper Entish

Standard Latin alphabet is used for creating names; i.e., letters, digits, and _ for short names (for example *agent_1*) and also: . / : # for complete names (URIs), for example, *pegaz://ipipan.waw.pl/DictionaryService#fun_1* We also reserve the symbols: *" "* & for special purpose, i.e., for constructing the elements of types *Token, Time* in pEntish.

We distinguish the following classes of Entish names.

- URI is the class (space) of universal and unique names in our system.
- VAR is the class of variables; any string consisting of letters and digits with prefix *?* is recognized as variable, for example, *?x, ?y, ?z, ?x1,* and so on.

- FUN (subclass of URI) is the class of function names.
- REL (subclass of URI) is the class of relation names.
- TYPE (subclass of URI) is the class of type names.
- OBJECT (subclass of URI) is the class of object names.

For logic operators we reserve the following symbols / names.

- *not* is the negation, *and*, *or*, *implies* are respectively conjunction, disjunction, and implication operators whereas *true* is the atomic formula always true.
- the symbol = denotes definition equality; $a = b$ means that a represents (is defined by) b,
- the symbol *:* is for typing; *Soul: ?x* denotes that the variable *?x* is of type *Soul*, whereas *Resource: latex_file)* denotes that the object *latex_file* is of type *Resource.*

Parenthesis *( )* are used in the standard way. For convenience, the first letter of any element of class TYPE is in uppercase.

Suppose that the name *go* belongs to the class FUN, and *Typ3: go(Typ1: , Typ2:)* is its signature, that is, the first variable of the function *go* is of type *Typ1*, the second variable is of type *Typ2*, whereas the value is of type *Typ3*.

If object *c1* is of type *Typ1* and object *c2* is of type *Typ2* (formally, *Typ1: c1* and *Typ2: c2*), then the the term *go(c1, c2)* is of type *Typ3* , formally
Typ3: go(c1, c2).

In Entish, new functions can be defined in the following way. If *put* is an element of FUN,
Typ2: put(Typ4: , Typ5:) and *Typ4: c4*, then
Typ3: new_fun(Typ1: ?x, Typ5: ?v) = go(?x, put(c4, ?v))
is the definition of function *new_fun* that has the following signature:
Typ3: new_fun1 (Typ1: , Typ5:)

Once *new_fun* is defined, it becomes an element of FUN. Let as note that in Entish a user can define a new function not necessarily from the functions already belonging to FUN. The user can also introduce new primitive functions using the definition schema above in the following way.
Typ3: primitive_fun(Typ1: , Typ5:) = // commentary in e.g., a natural language */*
In fact, the user introducing the new primitive function *primitive_fun* declares only the signature of the function with optional commentary on the right side of the definition equality symbol. Once the new primitive function is defined, it becomes an element of FUN.

In the very similar fashion we introduce to Entish, via definition, new relations composed of already existing ones (belonging to REL) as well as new primitive relations. Both composed and primitive relations can be introduced by any user and become elements of REL.

In the analogous way new types can be introduced to Entish. The new type names become elements of the class TYPE. For example, the following definition
TYPE: Nov_typ = // a commentary in a natural language */*
introduces the new type *Nov_typ* to our language. However, to be useful, new functions that operate on this new type should be also introduced (defined).

The problem is how to realize the distributed introduction of new primitive and composed functions, relations, and types to Entish. The solution to that problem is extremely simple and follows the idea of T. Berners-Lee of webizing language [8]. Any definition must be written down in a document (or in a service) that has a URI. In our system the service is called DictionaryService. The service URI concatenated with short name of the defined item gives out the URI (the complete name) of the defined item. Suppose that the URI of the DictionaryService, where the function *primitive_fun* was defined, is *pegaz://ipipan.waw.pl/DictionaryService* ; here the term *pegaz* refers to a specific protocol (a communication platform) on which the agentspace was built. Then, the complete unique name of *primitive_fun* is the following URI:
pegaz://ipipan.waw.pl/DictionaryService#primitive_fun

The Entish syntax resembles the syntax of KIF. So that an atomic formula is written in the following form: *(pred, c1, ?y)* where *pred* is an element of class REL, and its signature is the following: *(pred, Typ1: , Typ3:).* It means that the relation *pred* has two variables; the first one is of type *Typ1* whereas the second one is of type *Typ2* . In the example above the object (term) *c1* is of type *Typ1*.

The construction of the class of terms as well as the class of formulas is the standard one. Any element of class OBJECT is a term. Any variable is a term. If *fun* belongs to FUN and *Typ3: fun(Typ1:, Typ2:)* is its signature and *Typ1: term1* and *Typ2: term2*, then
fun(term1, term2) is a term of type *Typ3* .

Also the construction of the class of formulas is the standard one. The atomic formulas are of the following form: *(rel, ter1, ter2, ... , terN)* for any *rel* belonging to REL and having the signature
(rel, Typ1:, Typ2:, ... , TypN:) ,
and *Typ1: ter1, Typ2: ter2, ... , TypN: terN* .

If *form1, form2* are formulas, then
form1 and form2,
form1 or form2,
not form1,
form1 implies form2
are also formulas.

The language defined above is a simple version of the language of first order logic with types. The language was **webized** so that it is open and can be used and developed in a distributed way by users.

9.2 Proper Entish

Proper Entish is an instance of uEntish where specific types, relation and function names, as well as atomic formulas are introduced.

Let us start with type names. We introduce the following type names that belong to the class TYPE: *Agent, Node, Service, Soul, Operation_type, Operation, Resource, Time, and Token.*

For any *ag* of type *Agent* we introduce to Entish the following atomic formulas:
knows(ag),
goals(ag),
intentions(ag).
We introduce also the term *commitments(ag)* that is of type *Operation_type*. They should be seen as agent mental attributes. We also introduce the following attribute: *soul(ag1)*. It is an element of the type *Soul* and denotes the soul of the agent *ag1*.

In the analogous way we introduce to Entish the following formulas: For any *serv* of type *Service* : *intentions(serv)* is an atomic formula.

For any *serv* of type *Service* and any *oper* of type *Operation* the following:
commitments(serv)
type_of_operation(serv)
type_of_operation(oper)
are terms of type *Operation_type*. Operation type of a service expresses what is performed by the service, i.e., what is input and what is output of the service. Operation type of operation expresses what will be performed if this operation is implemented as a workflow. Let us note that the syntactic form of commitment and operation type is the same, however their intended meaning is different. Commitment consists also of two Entish formulas: precondition and post condition. The precondition describes input whereas the post condition describes the output of the operation performed by the service. In the case of commitment, the postcondition is what a service (agent) commits to realize, if the precondition is satisfied. The main difference between operation type and commitment is that usually a time-out is added to the commitment precondition and post condition.

For any *op_typ* of type *Operation_type* the following:
form_in(op_typ)
form_out(op_typ)
are atomic formulas introduced to pEntish. The formula *form_in* is the precondition of an operation type (commitment), whereas *form_out* is the post condition of an operation type (commitment).

For any element *soul* of type *Soul* we introduce the following attribute:
agent(soul); it is an element of type *Agent*. Its intended meaning is that it is the name of the agent to whom the *soul* belongs to. The introduction of the type *Soul* as well as its attribute serves for implementing "the soul migration" in our framework.

The type *Resource* contains all objects (except agents) that can be moved form one place to another. The idea is that the elements of this type are to be processed by workflows constructed and managed by the agents. We define the following attribute of object of type *Resource*. For any *res* of type *Resource* the following *token(res)* is an element of type *Token*. A token of a resource is the resource identifier in a workflow. The identifier must be locally unique for the service to which it should be delivered. Token is implemented as string, e.g., *"123abc"*.

We introduce the following relation symbols to pEntish.

– = denotes polymorphic equality relation for two terms of the same time: the intended meaning of the formula *(=, term1, term2)* is that the terms *ter1, term2* are of the same type and *ter1* equals *term2* .

- *is_in* denotes the polymorphic membership relation that has one of the following signatures:
  (is_in, Resource: , Service:) it means that resource is in service,
  (is_in, Agent: , Node:) it means that agent is in node,
  (is_in, Service: , Node:) it means that service is in node.
- *leq* denotes the inequality relation defined on the type *Time*. The relation has the following signature *(leq, Time: , Time:)* .
  The formula *(leq, date1 , date2)* denotes that *date1* is less or equal to (not later than) *date2*. This formula represents the general form of timeout in our framework.

We introduce to Entish the following format for element of type *Time* :
it is a string of the form *"hh:nn:ss/mm/dd/yy"* where *h,n,s,m, d, y* are digits, and denote respectively hour, minute, second, month, day, and year.

We introduce also global function denoted by *gmt*, that is necessary to express timeouts. It signature is *Time: gmt()* . The function has no variables, however its value, i.e., *gmt()* expresses the current GMT time on the local node. Let us note that this function is evaluated by agent perception. Since the agentspace is built on the top of Internet, GMT time is available on any host.

The Proper Entish introduced above constitutes the core of our approach to service integration. pEntish is rich enough to express static situations concerning agents, their goals, commitments, intentions and knowledge as well as the commitments and intentions of services, and relations between resources, places (nodes), services and agents. Also timeouts for commitments can be expressed here. Let us note that commitments serve as means for expressing agreements, cooperation between agent and service. A commitment consists of two Entish formulas: precondition and post condition. Their intended meaning is the following: service commits to realize the post condition (usually an intention of an agent) if the precondition is true. It is the job of the agent to make the precondition true. Usually, the precondition becomes the next intention of the agent. Precondition as well as postcondition is a conjunction of a timeout and a formula expressing that a certain resoutce is delivered as input to a certain service.

Let us note that there are no actions in our language. So that also causal relations can not be expressed here. Our language was design to be purely declarative where only static situations can be described. The intended semantics of agentspace, that is, the rules of agentspace functioning as well as agent and service behavior, must be expressed formally in the event structures. The event structures are used to define formal semantics of temporal languages with actions. In such languages agent / service attributes are viewed as modalities. Informal specification of agent / service behavour is presentend in Section 9.4.

9.3 Communication Entish

cEntish serves for constructing conversation policies as well as for transporting resources. It consists of specifications to be implemented as the standard in agentspace. It is important to stress that pEntish was design to describe situations (as formulas) in agentspace so that it may be viewed as a contents language. However, cEntish provides envelopes for message exchange in the course of workflow construction and management, as well as for transporting resources during workflow execution.

cEntish is to be interpreted by agent mind and services. Generally, cEntish consists of the specification of several data structures, to be implemented in XML, and a proposal of a universal conversation policy.

In cEntish we distinguish only the following data structures (kinds for short): `Control`, `Info`, `Soul`, `Operation`, and `Resource`. The pEntish formulas are used as attributes only in the following kinds: `Info, Soul, Operation.`

The first kind is `Message`, that specifies the general structure of cEntish message. Let `message` be of kind `Message`. Then, `message` has the following attributes:

- `id( message )` it denotes message identifier of kind `Id`.
- `from( message )` it denotes URI of agent or service.
- `to( message )` it denotes URI of agent or service.
- `contentsKind( message )` it is the name of kind of the resource included in the contents.
- `contents( message )` it is an instance of `Control, Info, Soul, Operation`, or `Resource.`

`Id` is the kind of message identifiers, it consists of strings of digits and letters.

The kind `Info` is design to express facts about agentspace. Let us define the format of data of kind `Info`. Let `info` be of kind `Info`. The attributes of `info` are the following:

- `formula( info )` it is a pEntish formula that describes the situation the `info` is about.
- `time( info )` it is GMT time of creating `info`, expressed in the string format of the type *Time*.
- `place( info )` is the URI of the place where the `info` was created.
- `from( info )` is the URI of the one (agent / service) who has created the `info`.
- `to( info )` is empty string or the URI of the one (agent / service) to whom the `info` is to be sent.

The format of `op` of kind `Operation` is defined as follows. It consists of the attributes:

- `operationType( op )` it is a pair of pEntish formulas.
- `algorithm( op )` it is a list of pEntish formulas; the intended meaning is that if agent adopts it as its plan, then the agent should realize the formulas as its intentions in the order given in the list.

This kind was designed to express the ways in which agent goals was (could be) realized.

The kind `Soul` is designed to implement our concept of agent mobility called *soul migration*. By the definition, it should contain all essential data of agent functioning. The format of kind `Soul` is defined as follows. Let `soul` be of kind `Soul`. We define the following attributes:

- agent(soul) - the name (URI) of the agent to whom the soul belongs to.
- goals(soul) - a list of pEntish formulas.
- commitments(soul) - a list of elements of the kind Commitment;
- intentions(soul) - a list of pEntish formulas; the intended meaning is that the formulas should be realized in the order they are in the list. If the formula is realized, it is removed from the current intentions, and saved in the history.
- knowledge(soul) - a list of elements of kind Info.
- history(soul) - a list of pEntish formulas (the intentions that have been already realized in the chronological order).

The kind Commitment serves for storing agent / service commitments. A commitment is a form of agreement or contract between two sites: the first site initiates the commitments and commits to realize some formula called postcondition, if the second side realizes another formula called precondition. An instance of Commitment represents the agreement, and it is stored in the soul of the first site. THe first site is obliged to notifiy the second site about the realization of the commitment. The kind Commitment has the following three attributes:

- form_in(commitment) it is an Entish formula that corresponds to *form_in* (i.e., precondition of the commitment);
- form_out(commitment) it is an Entish formula that corresponds to *form_out* (i.e., postcondition of the commitment);
- for(commitment) it is the full name (i.e., the addres URI) of agent or service (i.e., the second site) for whom the commitment was made.

triples where the first two elements are of pEntish formulas; the first formula of a pair corresponds to *form_in* (i.e., precondition) whereas the second one to *form_out* (i.e., post condition) of commitment.

The kind Resource serves as container for resources that are to be processed by services, and are not interpretable by agents. The format of the kind Resource is defined as follows. We introduce the following attributes: for any reso of kind Resource:

- token(reso) is a string of the type *Token* already defined in Section 9.2;
- type_of(reso) it is the name of type of the resource;
- data(reso) it is data.

The reason for introducing the kind Resource is to distinguish between data interpretable by agent mind (i.e., Soul, Info, Operation, Control, Id) and those that are not interpretable, i.e., being of kind Resource. The more important reason is that the elements of the kind Resource are processed in the workflows constructed and managed by agents.

The kind Control is defined as follows. It consists of the following strings: INI to initiate a conversation, OK to agree to start a conversation, and NOK to disagree. Although the format of Message, Soul, Info, Operation, Resource) seems to be fixed at the current stage of the project, perhaps the format of Control, Id should be extended to make workflow management more easy and explicit.

The design of the all Entish machinery presented above has the one main purpose: realizing the idea of simple protocols for constructing, executing and managing workflows by agents. The very protocols are the subject of the next section.

Universal Agent - Service Conversation Protocol. Let *ag1* be the name of agent, whereas *serv* be the name of a service. Agent *ag1* always sends to any service *serv* a message containing `info` with the intention formula of the following form:
intentions(ag1) implies form1 . The intended meaning of the intention formula is that the formula *form1* describes a class situations the agent wants to be in one of them.

If the the service *serv* is in an ordinary service performing some function on resources, and it is alble to realize the agent's intention, and it is ready to do so, then the service replays with the commitment formula that is of the following form:
(form_out(commitments(serv)) implies form1)
and
(form2 implies form_in(commitments(serv))

The intended meaning of the commitment formula is that the service *serv* commits to realize (to make true) the formula *form1* if the formula *form2* is realized. It is a job of agent *ag1* to realize the formula *form2*. In this way the service make commitment towards the agent to perform a job.

Any service that has committed towards an agent) is obliged to sent to the agent a notification (either positive or negative) in the form of info containing a formula from which follows either the formula *form1* or its negation, i.e., *not form1*. In the current (still experimental) version of Entish it is not clear if this form of notification is sufficient. Perhaps a special form of notification should be introduced.

If, however, the service *serv* is an InfoService, then, usually, the service sends back the formula of the following form: *form_out(cos) implies intentions(ag1)* .

The intended meaning of this formula is that *cos* is the address (complete name) of a service that can realize the intention of agent *ag1*, or *cos* is the address (complete name) of operation that, if adopted as plan, could help the agent to realize its intention.

The conversation protocol defined above is general so that it can be applied in various domains. However, we are interested only in workflow construction and management. Generally, a workflow consists in scheduling resource processing by services. That is, resources of appropriate types should be delivered (as input) to a service (participating in the workflow) in time, then the output (i.e., the resource being the result of processing the input by the service) should be delivered to the next scheduled service of the workflow. For this very purpose, *form1* in the intention formula is always of the following form:
(is_in, reso_k, serv_k) and (=, token(reso_k), string_k) and (leq, gmt(), time0) . The intended meaning is the following: the resource *reso_k* with the token *token(reso_k)* is in the service *serv_k* by the time *time0*.

In the commitment formula, *form2* has always the following form:
(is_in, reso_k1, serv) and (=, token(reso_k1), string_k1) and
(is_in, reso_k2, serv) and (=, token(reso_k2), string_k2) and ...
(is_in, reso_kN, serv) and (=, token(reso_kN), string_kN) and
(leq, gmt(), time1)

Where *N* is the number of input resources the service *serv* needs in order to produce the resource *reso_k*. If *N=0*, then the formula *form2* is of the following form: *true and (leq, gmt(), time1)* . Finally, let us note that *(leq, gmt(), time1)* is a timeout set by the service *serv* for performing job requested by the agent *ag1*. Let us note that a service, that has already committed to perform a job for an agent, expects to receive resources having the tokens precisely defined.

Let us explain how the conversation protocol works on the example of BodyService. Operation type of BodyService located at the place *place1* is defined as follows. The formula
form_in(type_of_operation(BodyService))
is represented by the following:
(Soul: ?x) and (is_in, ?x, BodyService) and
(is_in, BodyService, place1) .
Whereas the formula
form_out(type_of_operation(BodyService))
is represented by the following:
(is_in, agent(?x), place1)) .
The intended meaning may be expressed as follows. If the soul *?x* is send to the *BodyService* located at the *place1*, then the agent *agent(?x)* with the name the same as the name contained in the soul *?x* is in the *place1*.

Suppose that the agent *ag1* wants to migrate to the place *place1*. The agent sents to the *BodyService* at the *place1* the info containing the intention formula:
intentions(ag1) implies (is_in, ag1, place1) .
Then, if the service is ready, it sends back the following commitment formula:
((is_in, soul(ag1), BodyService) and
(leg, gmt(), "09:00:00/09/07/01")) implies
form_in(commitments(BodyService)))
and
(form_out(commitments(BodyService)) implies
((is_in, ag1, place1) and
(leg, gmt(), "09:00:01/09/07/01"))) .
The intended meaning is that if agent *ag1* sends its soul to *BodyService* by 09:00:00/09/07/01 GMT, then the agent will be moved to *place1* by 09:00:01/09/07/01 GMT. However, in order to exclude the situation where two copies of the same agent exist at the same time, the agent must close its process on the old place immediately after sending its soul to the new place.

9.4 Specification of Intended Agent / Service Behavior

The following list of conditions specifying the desired behavior agents and services presented below is certainly not completed.

- A service is allowed to make a commitment in response to agent intention, only if it can realize the job requested by the agent. The service should do it best to realize the commitment before the timeout. The service is obliged to sent notification about commitment realization.
- There can not be two copies of the same object at the same time. It is of particular importance in the case of agent migration where agent's soul is moved to another place. If the agent process is not closed at the old place, then two copies of the same agent are in the system, and each of them is autonomous. This may cause serious problems because they may disturb each other.

- Agent's goals and commitments can not be canceled by the agent. Agent's intention can not contradict its goal or commitment.

9.5 Implementations Details

First of all, the space of URI should be defined. Analogously to URL, we propose the following protocol for creating URI in agentspace:
pegaz://ii-ap.siedlce.pl/node/service#name
where *pegaz* corresponds to a communication platform (protocol).
ii-ap.siedlce.pl is a host name.
node is the name of place on the communication platform.
service is name of service running on this node.
name is name of object, agent, or defined notion located at this service.

An important point in our framework is soul creation. The soul creation is reserved exclusively to a SecretaryService. SecretaryService is a GUI to agentspace. There are two special services associated to any SecretaryService, i.e., MailService and DictionaryService. A SecretaryService is also the place where the agent is given its complete name (URI) as an attribute of its soul. Let us note, that in our framework, agent complete name is at the same time the mail address of the agent for asynchronous communication. Of course, agent being mobile, i.e., moving from place to place, communicates immediately from its local place. However, if a service wants to communicate with such mobile agent, it uses the agent name as the permanent static address of the agent. Mail account are located on MailService that must be ran on a persistent server. The messages sent to agent mail account may contain intermediate resources of the workflow that are sent to the workflow owner (i.e., the agent) in the case of a partial failure of the workflow execution. Notifications of realizing commitments are also sent to the agent mail account.

DictionaryService implements Entish syntax, Here, a user can define its new concepts, names, and symbols according to the Entish rules. The implementation of DictionaryService is left to a user. The only requirement is that the document where new definitions are written down, must be visible (as read-only file) from the outside like the documents on www. The format of DictionaryService document must be fixed and implemented in XML. This will be done shortly.

At the current stage of the project, the data structure (kinds) specified in cEntish are implemented in Java. Since XML is a standard, the final version of the data structure will be implemented in XML.

Toward a Multi-agent Modelling Approach for Urban Public Transportation Systems

Flavien Balbo[1,2] and Suzanne Pinson[2]

[1]Inrets – Gretia
2, avenue du Général Malleret-Joinville, F-94114 Arcueil cedex
[2]Lamsade, Université Paris-Dauphine
Place du Maréchal de Lattre de Tassigny 75775 Paris cedex 16
balbo@lamsade.dauphine.fr
pinson@lamsade.dauphine.fr

Abstract. In this paper, a multi-agent system (MAS) for bus transportation management is presented. The aim of our MAS is 1) to diagnose problems in the bus lines (bus delays, bus advances,...) and 2) to detect inconsistency in positioning data sent by buses to the central operator. Our MAS behaves as a Multi-Agent Decision Support System (MADSS) used by human regulators in order to manage bus lines. In our model, buses and stops are modeled as autonomous agents that cooperate to detect faults (disturbances) in the transportation network. An original interaction model called ESAC (Environment as Active Communication Support) was designed to allow non-intentional as well as direct communication. The system was implemented using ILOG RULES and was tested on data coming from the Brussels bus transportation network (STIB).

Keywords: Multi-agent system, Decision Support System, interaction, environment, dynamic systems, transportation system.

1 Introduction

Designing, implementing, and adjusting urban public traffic control systems involves quite effort and knowledge. The effectiveness of urban traffic control systems greatly depends on its ability to react upon changes in traffic patterns (for example traffic jams, road work, one-way streets, passenger clusters, etc.). The hypothesis is that it is useful for human regulators (the staff in charge of monitoring the network) to design Transportation Decision Support Systems that is able to adjust itself if the environment changes, that is to say to automatically detect incoherent data, traffic disturbances and then to automatically propose solutions to optimize the traffic flow in order to maximize the person and vehicular throughput and to minimize delay.

This research is part of the SATIR project done with the participation of the French Transportation Institute INRETS (Institut National d'Etudes et de Recherche sur les Transports).

A. Omicini, P. Petta, and R. Tolksdorf (Eds.): ESAW 2001, LNAI 2203, pp. 160-174, 2001.

Urban public transportation systems are naturally open systems (vehicles appear in or disappear from the network according to their activity) and distributed systems (vehicles move on a network).The multi-agent paradigm makes it possible to model those systems where the distribution of control and knowledge facilitates problem solving, improves robustness and reduces execution time. Therefore, a multi-agent approach was chosen to model the system in order to 1) diagnose disturbances in the bus lines (bus delays, bus advances,...), 2) detect inconsistency in positioning data sent by buses to the central regulator,3) dynamically compute schedules and 4) monitor and process disturbances. In this paper, we will only present the two first points. The last ones are part of the whole SATIR project [2].We applied our approach to a bus transportation network.

A variety of approaches have been used to model urban transportation network (Operation Research methods, fuzzy theory, neural networks, genetic algorithms, intelligent agents). A good review of the literature can be found in [2,10]. Very few models are based on the multi-agent paradigm because one of the difficulties in both the design and the understanding of MASs comes from the lack of central controls and the ensuing conflicting, uncertain, incomplete and delayed knowledge on the part of the agents. LIN used the multi-agent approach for transportation scheduling and simulation in a railroad scenario [9]. Brezillon has designed a simulator to help human regulators of the Parisian subway [5]. Theses systems present several drawbacks: they are mostly simulation systems that are not integrated into the decision support system and they are not directly fed with real-time data coming from vehicule sensors.

In the following, we present the global design of the multi-agent system (MAS) for bus transportation management.

The second section presents notions of the urban public transportation domain. In the third section, the overall architecture of our system is presented. In our model, buses and stops are modeled as autonomous agents that cooperate to detect faults (disturbances) in the transportation network. An original interaction model called ESAC (Environment as Active Communication Support), presented in the second section, was designed to allow non-intentional as well as direct communications. The last sections show how the system operates and present the implementation of the system and its validation on data coming from the Brussel bus transportation network (STIB).

2 Notion of the Domain

In urban transportation control, human regulators are located in a control center. They have to manage the transportation network under normal operating conditions (where are the buses located ?) and also under disturbed conditions (where are disturbances (bus delays, bus advances) located ? What action has to be taken to solve the problem?).

In most networks, vehicles are located through sensors which provide real-time information. This information represents a huge amount of data (data arrives every

40 seconds). Furthermore it may be incomplete (a sensor breaks down) or uncertain (the quality of the data is sometimes poor). This data is collected through the Automatic Vehicle Monitoring systems (AVM). AVM compares the actual positions of the vehicles (captured by the sensors) with their theoretical positions in order to provide the regulator (the staff in charge of monitoring the network) with an overview of the routes. In this way the regulator can see whether the vehicles are running ahead of schedule or are running late. .

Figure 1 shows the AVM management of real-time information coming from sensors and the output of the system. Each line is represented two ways with its stops and its running buses. Each bus location is represented by 1) a number for its theoretical position coming from theoretical schedule, 2) a colored square for its real location detected by the system. This real location may be erroneous due to sensor break downs. Stops are represented by black dots. The gap between the theoretical position and the real position gives an information about the bus delays or advances. Colors give the importance of the delays or advances.

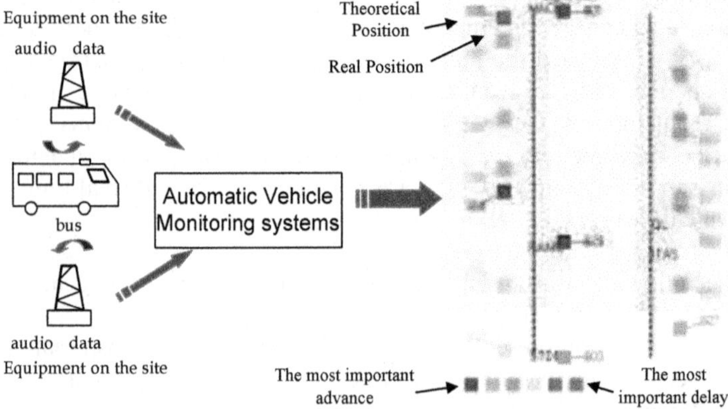

Figure 1. Real-Time Information in a Bus Transportation Network.

Problems arise when a vehicle may not be announced on stops he has served thus generating a false alarm (declared as late while it passed over the stop) and erroneous output on the screen. We will see how our model allows to manage these inconsistencies without using broadcasting.

3 The Overall Multi-agent Model

The agent technology is based upon the notion of reactive, autonomous, proactive entities that inhabit dynamic environment. Multi-Agent Systems can be characterized by the interaction of many agents trying to solve a problem in a cooperative way. Several methodologies were proposed to model Multi-Agent System (MAS). Our approach is based on the AEIO methodology ([6]). Following this approach, we defined an Agent model, an Environment model and an Interaction model. The organization model, needed to find a solution to the detected disturbance, is defined in another part of the project and is described in [3, 11]. This paper focuses on the AEI part of the model, principally on the Interaction model that we have called the ESAC model.

After an analysis of the problem, we defined two types of agents : STOP agents and BUS agents (cf. fig 2). Each STOP agent calculates a schedule for a Bus to come on its line at its position and the BUS agents account for the activity of the vehicle by announcing when they will arrive at the stop concerned. A STOP agent can trigger an alarm if an expected BUS has not been announced while its schedule is over.

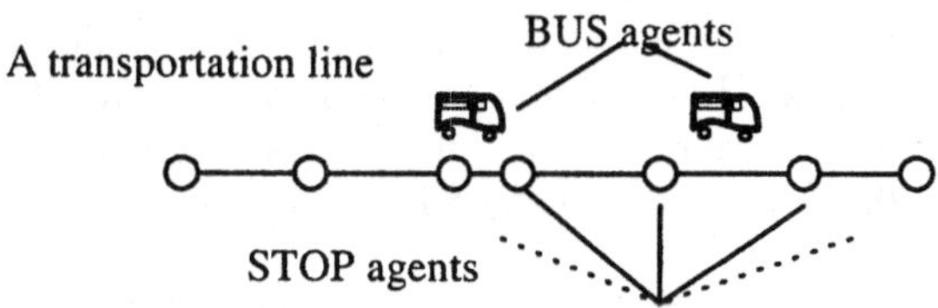

Fig 2. A Multi-agent Urban Public Transportation Line.

The ESAC model is a communication model between agents based on the environment as an active communication support. It differs from the four most widely known interaction models used in MAS: a sender broadcasts its messages, a sender knows the name of its receivers, a sender uses the Contract Net Protocol, a sender send messages through middle agents. In urban traffic control domain, the sender does not always know the name of its receivers. For example, when a bus has to contact its nearest bus, it does not know its identification. We designed an original communication model, ESAC, that addresses this type of problem. This leads to a type of agent model that encompasses the characteristics of the receivers and that makes this knowledge accessible to the environment. The use of the environment as active communication support also supposes that the sender needs are represented in the structure of the messages. This section introduces the ESAC model and its various components.

3.1 ESAC: An Agent Model

The aims of our agent model are twofold :

- To allow the environment to identify the characteristics of an agent
- To give the agent the knowledge necessary to participate in the environment.

A general overview of the components of an agent is given below, followed by the description of two original modules: the **Public Layer** module and the **Environment Knowledge** module.

In our system, the agents can be heterogeneous and can belong to different categories (for example, the STOP agents and the BUS agents).

The agent model that we propose is made up of five modules:

- The **Communication** module, which enables the agent to send and receive messages.
- The **Domain Knowledge** module, which contains the knowledge necessary for the agent to work.
- The **Control** module, which manages the interaction between the modules. The information flows between the modules are directed by the Control module.
- The **Public Layer** module, which contains all the information that is accessible to the environment.
- The **Environment Knowledge** module, which contains the knowledge necessary for the agent to interact in the environment.

In the following, we are going to detail the last two modules. The communication module is the classical one found in every MAS. The domaine knowledge is dedicated to the application domain. It contains planification knowledge for BUS agents and the theoretical schedule for STOP agents.

3.1.1 The Public Layer Module

The information contained in this module is represented as pairs:

(attribute_name, attribute_value)

For example for a BUS agent:

(last_position, 10); (number, 54806); (line_number, 54).

It identifies each of the representatives from a category of agents and is accessible to the environment. The combination of this information enables the environment to identify a sub-set of representatives from a category of agents, for example all the STOP agents situated between the active bus position and the terminus stop.

3.1.2 The Environment Knowledge Module

In order to enable the agents to receive messages, we define communication filters. These filters constitute the **sensors** of the agents on their environment.

The knowledge held in the Environment Knowledge module is necessary for an agent to receive messages. It defines the appropriate receiver. For example, a STOP agent may want to contact all the BUS agents that are going to stop. This module is divided into two parts, the first one containing the communication filters which allow him to receive the messages by pattern-matching, the second one concerning the social knowledge of the agent.

- **Communication filter module**

We define three types of filters: the emission filters, the reception filters and the interception filters. They are detailed in 3.3.

- **Social module.**

This module contains two types of information:

- ***Category acquaintances***, which concern the description of the Public Layer module of the categories of agents with which the agent may communicate. For example, the category "STOP agent" has the description of the public layer of the category "BUS agent".

- ***Special acquaintances***, which contain the information that allows an agent to send messages to the agents with which it has privileged relations. For example, a STOP agent has privileged relations with the preceeding STOP agent and the next STOP agent. This information is used by the agents when a privileged relation between agents cannot be represented logically with filters.

3.2 ESAC: A Communication Language

We propose a communication language which is common to all of the agents in the system (for a detailed description, see [1; 4]). Every message sent or received has to comply with our communication model in order to enable the sender to express his needs and to be understood by all of the agents.

Messages have the following structure:

<message>::=<environment> <reception> <body>

In the following paragraphs, the elements which make up the <environment> and <reception> parts of a message are presented. The <body> part will not be presented because it based on the performatives which are traditionally used in high level languages (KQML [7], FIPA ACL [8])

3.2.1 The Structure of a Message: The Environment Parameters

<environment>::=<validity_limit><priority> <receiver_category>

The information referred to in this part is the data which allows the environment to process the message. The parameters are meta-information which is not directly related to the body of the message (containing domain knowledge) but to the way in which it should be processed by the environment.

Currently three types of information are used:

- **<validity_limit>**: a sender agent puts its query into the environment without necessarily knowing the receiver. There may be just one receiver but the message may also concern several receivers. In this case, the message must remain available for the whole community. We have chosen the following convention: every message sent to a single receiver with a single identifier or as the result of a comparison between the potential receivers, is destroyed by this receiver. A message with several potential receivers is destroyed by the environment (according to the value of <validity_limit>).
- **<priority>**: the value of this attribute represents the importance of the message. It is given by the sender according to its needs. It is also used to determine the importance of the filter which will process the message. The environment sorts these messages according to their priority, thus taking into account the needs of the agents.
- **<receiver_category>**: an agent indicates to the environment the agent category to which its message is sent by attributing values to this parameter.

3.2.2 The Structure of a Message: Reception Conditions

<reception>::=<filter_name > (<condition>)

The second part of the information corresponds to the data used in order to find the receiver of a message. The first characteristic, called **<filter_name>**, indicates the name of the filter.

The second attribute **(<condition>)** puts together the values which will be used to search for the receiver, in a post -fixed polish notation (the operator is written at the end of the mathematical or logical expression, and the conjunction of predicates is between parenthesis)

((<attribute1_name, attribute1_value, operator1>)

...

(<attributeN_name, attributeN_value, operatorN>))

operatorI is either
- a mathematical comparator.
- an application dependent predicates. For exemple, ">FirstPosition" is used to locate the closest BUS to a given position. In the pattern-matching process, the environment uses this predicate to compare potential receiver agents (cf. Section 3.3).

In every triplet, the operator *operatorI* compares the *attributI_value* contained in the Public Layer module of the potential receiver to *attributI_value* contained in the message.

For instance:

(position, 10,>)⇒ position_of_potential_receiver> 10

This triplet means that the value of the position attribute of the receiver (held in its Public Layer module) must be over 10.

In this way, we define "a communication logic" enabling each agent to find its interlocutor according to the characteristics it is searching for in this agent.

3.3 An Active Environment

The environment is made up of the reception and interception filters and of the representatives of each categories of agents. When an agent wants to communicate, its puts its message in the environment. The pattern-matching process consists of three steps. In the first step, the environement uses the information <receiver_category> to take into account a set of filters and the category of the agents concerned with the communication. The filter name (contained in the message) is used during the second step to match the filter to take into account. During the third step the selected filter is instantiated with the attribute values of the filed <*condition*> contained in the message. This step makes it possible, through the application of the instantiated filter, to choose the receiver agents concerned with the communication.

Three types of filters have been defined in our model. Each filter corresponds to a precise communication need:

- **Reception filters**: communication is based on a need that is common to two agents. The sender specifies the values of the characteristics searched for in the receiver. This description is matched against a communication filter of the category of the agents contacted.

The structure of a filter is described in figure 3:

(:filter =<filter_name>
:priority = <message_priority> + [<parameter>]
:condition = ((<attribut_name >, x, <operatorI>)*))

Figure 3. The Structure of a Filter.

The architecture of a filter is made up of three distinct elements:

- *<filter_name>*: this information identifies the filter from among the set of filters of a category of agents. The comparison of this value with the *<filter_name>* information contained in the message allows the environment to choose the filter.
- *<message_priority>*: this information is used to defined the order in which messages are processed (cf. Section 3.2.1). Each filter receives priority number from the *<priority>* field of the processed message; it may be modified by the optional parameter [*<parameter>*]. *<parameter>* modifies initial priority coming from a message.
- In the field *<condition>*, variables (x) are instantiated with the attribute values of the corresponding conditions in the message.

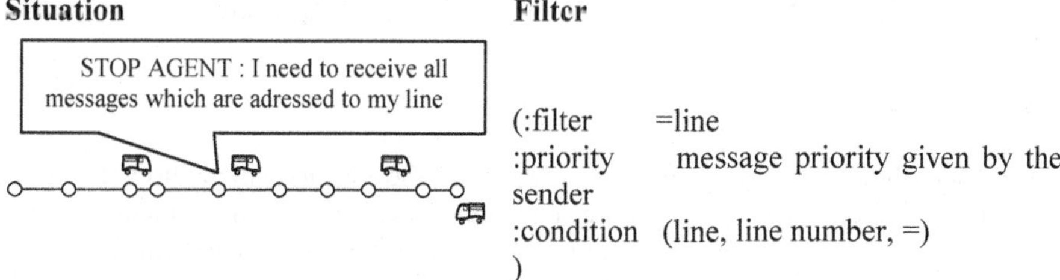
Situation
Filter
STOP AGENT : I need to receive all messages which are adressed to my line
(:filter =line
:priority message priority given by the sender
:condition (line, line number, =)
)

Fig 4. Example of a Reception Filter Operation.

With the reception filter, which name=*line* (cf. fig 4), an agent will receive all the messages sent to its line. This filter compares the attribute "line", located in the agents public layer, with the value contained in the message.

- **Emission filters**: communication is based on a need of the sender which doesn't match any expectation of the category of contacted agents. The consequence of this lack of interest among the receivers is the absence of suitable filters which would make it possible for the environment to define the agents concerned.

Consequently, the sender must put the appropriate filter in the environment at the same time as it sends its message. An emission filter may require the comparison of the potential receivers and therefore the use of predicates. This possibility implies that the filter matches all the potential receivers of a message and then searches for the exact receiver of that message.

```
(:filter              = <filter_name>
 :filter_priority     = <message_priority> + [<parameter>]
 :condition           = (((<attribut_nameI>, x, <operatorI>)*)
(¬((<attribut_nameJ>,< chosen_agent _value>,<operatorJ>)*) )))
```

Figure 5. The Structure of a Filter to Compare the Candidates.

In order to compare the candidates, conditions concerning the receivers are added to the initial structure of a filter. These conditions are identical to the ones used to define the candidates but the comparison values are those of the candidates and not those contained in the message.

For instance, the filter, which name is "*<FirstPosition.=line*" (cf. Figure 6), is used in order to find a particular BUS agent. It must be the closest BUS to the sender terminus. The sender expresses this need with a predicate : <FirstPosition. The receiver position must be less than the position value contained in the message. In the sub-set of the potential agents, the exact receiver must have the greatest value of attribut position.

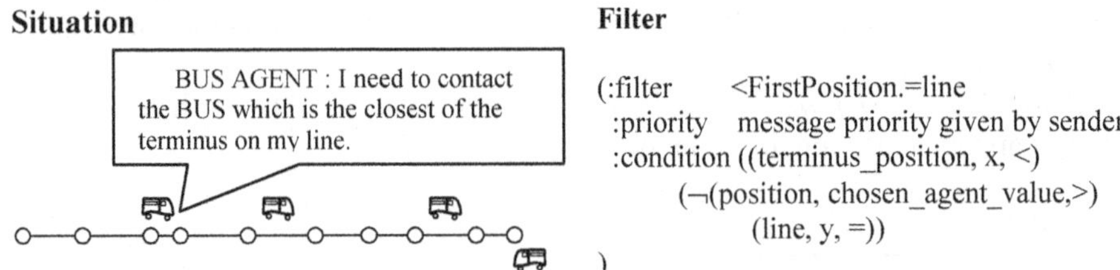
Situation
BUS AGENT : I need to contact the BUS which is the closest of the terminus on my line.
Filter
(:filter <FirstPosition.=line
:priority message priority given by sender
:condition ((terminus_position, x, <)
(¬(position, chosen_agent_value,>)
(line, y, =))
)

Fig 6. Example of an Emission Filter Operation.

- **Interception filters**: the interception filters are the sensors which allow the agents to receive the messages which are not applied to them but the content of which may be of interest for them. The goal of these filters is to make full use of the environment as a shared work context [12; 4] where every communication is potentially available for all of the participants.

The structure of an interception (cf. Figure 7) filter allows the environment to take into account the characteristics of the sender (through the body of the message), the receivers and the interceptor. There are therefore three types of conditions for an interception filter:

- ***The characteristics of the sender***: an interception filter includes conditions giving the characteristics of the message the interceptor wishes to receive (variable m).

- ***The characteristics of the receivers***: an interception filter includes in its field condition the reception filter which makes it possible to find the receivers of the initial message.
- ***The characteristics of the interceptor***: an interception filter includes in its structure some conditions governing the intercepting agent. The conditions refer to the attributes of the normal receiver (variable z) and/or of the message (variable m).

```
(:filter              = <filter_name>
 :priority            = <message_priority> + [<parameter_2>]
 :condition           =
      ((z(:filter              = <filter_name>
          :priority            =message priority + [<parameter_1>]
          :condition           = (((<attribut_nameI>, x,<operatorI>)*)
      (¬(<attribut_nameJ>, <chosen_agent_value>, <operatorJ>)*)))
       )
      (m (<information_Message>, y, <operatorK>)*)
        ((<attribut_Interceptor>, z, <operatorL>)*) ))
```

Figure 7. The Interception Filter.

In order for an agent to intercept a message, the interception filter must be given priority over that of the reception filter. The structure of a filter therefore includes information concerning its priority (<*priority*>) which is equal to the priority of the message (plus the value of [*parameter*]). Assigning to an interception filter the priority of the reception filter it includes plus 1 (at least) ensures that the environment will activate the interception filter before the communication filter.

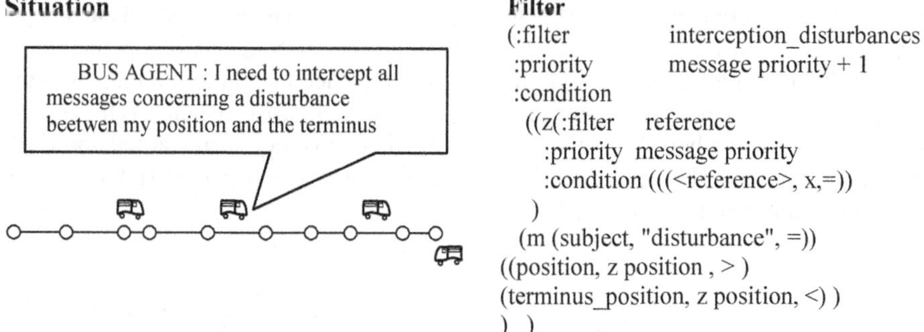

Fig 8. Example of an Interception Filter Operation.

For instance, the filter interception_disturbances (cf. Figure 8) is used to intercept all messages directly addressed to an agent (using the filter named "*reference*") and relative to a disturbance. The interceptor is interested by the message only if the position of the initial receiver is between it's own position and the terminus position.

This filter has a priority egal to the message priority plus 1 because the filter, named "*reference*", has a priority egal to the message priority.

The possibility of intercepting messages justifies the uses of filters as opposed to the automatic processing of the content of the various Public Layer modules. An agent indicates to the environment that it is interested in events (interaction between agents) which do not concern it directly. Automatic filtering of the content of the Public Layer of the category of receiving agents according to the conditions contained in the message doesn't allow this form of interaction. This original concept introduces a new type of interaction which doesn't emerge from the initial protagonists. The triggering of this interaction depends on the relations between the sender, the receivers and the interceptor.

4 An Application: The Management of a Bus Transportation Network

As we said before, the MAS was designed to manage the transportation network under normal operating conditions and also under disturbed conditions. Vehicles are located through sensors which provide real-time information. This data arrive every 40s which represents a huge amount of data. This data is collected through the Automatic Vehicle Monitoring systems (AVM). AVM compares the actual positions of the vehicles (captured by the sensors) with their theoretical positions in order to provide the regulator with an overview of the routes. In this way the regulator can see whether the vehicles are running ahead of schedule or are running late. Another problem arises when buses are not correcly located due to sensor dysfunctions (inconsistent date). The problem studied in this application is the management for the inconsistencies resulting from the data collected by sensors.

We have defined two categories of agents: (1) the STOP agents, (2) the BUS agents. They interact through messages inside the environment. The STOP agents have the timetables of the vehicles in service and the BUS agents report the actual state of the network by indicating to the STOP concerned (sending of a message) the passage of a physical bus at a physical stop. The problem arises when a vehicle is no longer located and when the STOPs on the bus route have not been warned about the passage of the bus. The consequence is that false alarms are triggered (a bus has not appeared and is running late) (cf. fig 9).

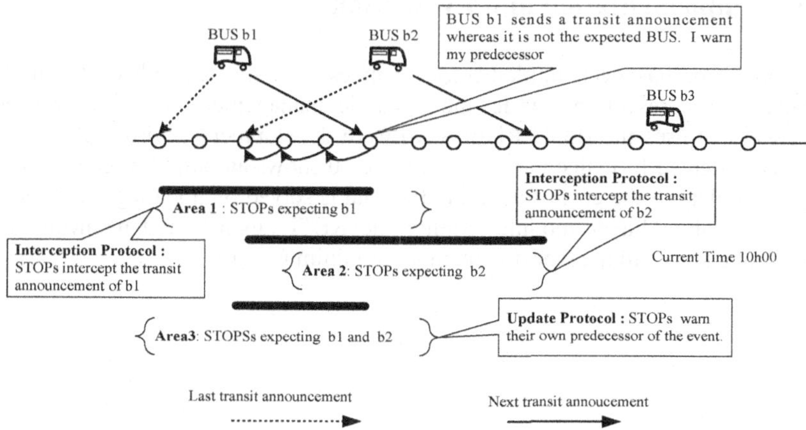

Figure 9. An Example of Incoherent Data.

In order to solve this problem, an interception filter can be used. The BUS cannot, by itself, inform the STOP about a stop it has gone past without stopping, because it has no knowledge about where it has supposed to stop. A STOP agent observing a delay cannot initiate the consultation of a BUS it is expecting for two reasons: (1) a stop can be passed by a vehicle running early and the STOP agent does observe any anomaly; (2) the BUS cannot answer because it doesn't know its route (by definition it doesn't know where it will be correctly located again).

The solution chosen is to intercept all new transit announcement sent by vehicles not running to schedule in order to make it possible for the STOPS concerned to receive the message and update their timetable.

- ***The filter takes into account the relation between the interceptor and the receivers***: out of all of the messages sent by the sender, the interceptor is only interested in those which have receivers with a position superior to its own.
- ***The filter takes into account the relation between the sender and the interceptor***: the filter intercepts the messages sent by the vehicle with a waiting interceptor.
- ***The filter takes into account the nature of the message***: the interceptor only wants to receive messages concerning a transit announcement.

The sender and subject information comes from the *<body>* part of the message (cf. Section 3.2).

This filter will be activated by the environment only when a vehicle has not been announced at the stops and when it "reappears" further on at a new stop. So, the STOPs that have not been informed receive the message from the BUS agent and update their schedule. The BUS and STOP agents interact through the communication according to the normal schedule updating protocol.

5 Implementation and Evaluation

A prototype has been implemented to investigate the feasibility of a multi-agent model to represent the static and dynamic management of o bus transportation network. It uses the ILOG RULES software development environment based on an object-oriented formalism and C++. Figure 10 shows an output screen of the system. The problem of the cost of updating the data is solved by updating the content of the agents' Public Layer module locally; the RETE algorithm dynamically manages content changes to the working memory at minimum cost.

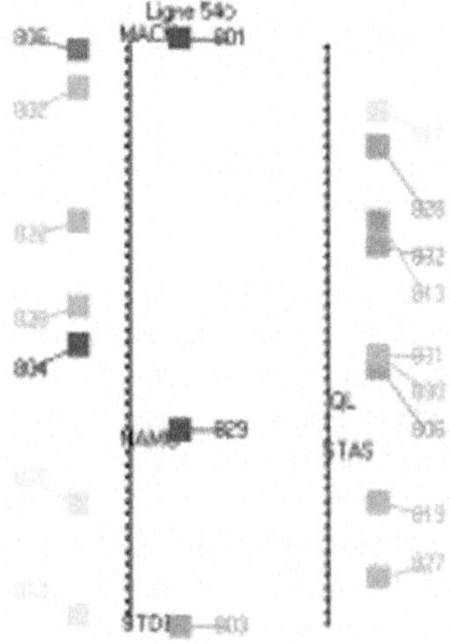

Figure 10. Output Screen of the MAS Public Transportation System (squares, dots and colors are explained in §2).

Two types of validation have been undertaken: face validity and predictive power. face validity is a surface or initial impression of the realism of a system and was checked in the early stages of model building. The system was shown to a group of experts (human regulators) of the Brussels Transportation Society (STIB). They considered the correspondence between the model operation and their perception of the real-life phenomena which the model represents to be good and so accepted face validity. Predictive power was evaluation by testing and comparing results of the system against reality. The assessment of the system covered 8 days of test of the recorded data covering two lines (lines 20 and 54) where more than 300 disturbances were identified. In parallel, the operators work were observed and the disturbance were precisely recorded at the control center. The tests show that our MAS detected all the disturbances while the operators detected only 10% of them. It can be explained by the fact that, when operators detect a disturbance, they immediately try to solve it, which is time consuming, preventing them to detect other disturbances.

It is obvious that more testing and validation are called for. Although the validation of the system is at its early stages, results so far have shown that the MAS approach offers potential to model bus transportation network in order to assist human operators to detect disturbances and that research should therefore proceed.

6 Conclusion

This paper shows how a multi-agent methodology and framework has been applied for the development of a distributed DSS for urban public transportation system management.

The aim of our MAS is 1) to diagnose disturbances in the bus lines (bus delays, bus advances,...), 2) to detect inconsistency in positioning data sent by buses to the central regulator and 3) to monitor and process disturbances. In this paper, we have only presented the two first points.

Several key contributions should be highlighted. First, multi-agent paradigm provides an adequate structure to represent multiple and complex interactions between agents. Second, the original concept of active environment that we have introduced, allows the agent to interact directly, sending and receiving messages through logical filters of emission, reception and interception. The technique of the active environment is used to make the model consistent when a bus has reappeared after several stops given its latest location. Interception logical filters enable the passed stops to update their next bus waiting time planning. Third, the multi-agent approach allows the dynamic management of the bus schedule, the bus schedule follow up in real time and the comparison with the theoretical bus schedule during the day . The disturbance detection and display enables the operators quickly to identify the malfunctions of the lines inside the network.

Our multi-agent system is used as the basis of a decision support system and has been assessed over several test days in real size. Using real data coming from the Brussels bus network, the system has been shown to work despite the sensors failures.

More research has to be done in several directions: 1) fully testing and validating the system, 2) improving the dynamic management of the filters, so that the agents can add and remove filters from the environment. This dynamic management requires the use of knowledge on category acquaintances. This new generation of agents is expected to increase its autonomy by building its own communication filters according to the evolution of its needs, 3) developing the user dialog and interfaces to allow human regulators to ask for explanations of the disturbances, 4) adding statistical tools to better understand the network operations, 5) taking into account several lines and the interchange problems.

References

[1] Balbo F., "A model of Environnement, active support of the communication", *Proceedings AAAI Workshop on Reasonning in Context for AI Application.* AAAI Technical Report WS-99-14, AAAI Press, Menlo Park, CA, Orlando 1999.

[2] Balbo F., "ESAC: un Modèle d'Interaction Multi-Agent utilisant l'Environnement comme Support Actif de Communication. Application à la gestion des Transports Urbains", *PhD Thesis in Computer Science*, Université Paris IX, January 2000.

[3] Balbo F., Scemama G., "Modélisation d'une perturbation sur un réseau de transport : le modèle Incident"*INFORSID* , Lyon, 16-19 mai 2000

[4] Balbo F., "The environment: a privileged intermediary for agent interaction", MAAMAW workshop (poster session), Annecy, France April 2001.

[5] Brézillon P., Pomerol J.C., "Contextual Knowledge and Proceduralized Context",. *Proceedings AAAI Workshop on Reasonning in Context for AI Application.* AAAI Technical Report WS-99-14, AAAI Press, Menlo Park, CA, Orlando 1999.

[6] Y. Demazeau, "From Cognitive Interactions to Collective Behaviour in Agent-Based Systems", European Conference on Cognitive Science, Saint-Malo, April 1995.

[7] Finin T., Fritzon R., McKay D., McEntire R., "KQML as Agent Communication Language", *CIKM'94*, ACM Press, 1994.

[8] FIPA 99 Specification Part 2, "*Agent Communication Language*", http : //www.cselt.stet.it/ufv/leonardo/fipa/index.htm.

[9] Lind J., Fischer K., "Transportation Scheduling And Simulation In A Railroad Scenario : A Multi-Agent Approach", DFKI Technical-Memo-1998-05.

[10] Niittymäki J., Pursula M., special issue : "Artificial Intelligence on transportation systems and science, European Journal of Operational Research, volume 131, n°2 2001.

[11] Scemama G., Balbo F., Caruso M., Rodriguez J., "Real Time Aid to Decision System for Bus Operators", 10th*International Conference on Road Transport Information and Control, IEE conference*, 4-6 April 2000

[12] Zouinar M., "Modélisation des Processus de Constitution du Contexte Partagée pour l'Analyse et la Conception des Environnements de Travail Coopératifs", Actes de *Ingénierie des Connaissances*, IC 98, Pont à Mousson, CNRS, 13-15 Mai 1998.

ITTALKS: An Application of Agents in the Semantic Web*

Filip Perich, Lalana Kagal, Harry Chen, Sovrin Tolia, Youyong Zou, Tim Finin, Anupam Joshi, Yun Peng, R. Scott Cost, and Charles Nicholas

Laboratory for Advanced Information Technology
University of Maryland Baltimore County, 1000 Hilltop Circle, Baltimore, MD 21250
{fperic1,lkagal1,hchen4,stolia1,yzou1,finin,joshi,peng,cost,nicholas}@csee.umbc.edu

Abstract. Effective use of the vast quantity of information now available on the web will require the use of "Semantic Web" markup languages such as the DARPA Agent Markup Language (DAML). Such languages will enable the automated gathering and processing of much information that is currently available but insufficiently utilized. Effectively, such languages will facilitate the integration of multi-agent systems with the existing information infrastructure. As part of our exploration of Semantic Web technology, and DAML in particular, we have constructed ITTALKS, a web-based system for automatic and intelligent notification of information technology talks. In this paper, we describe the ITTALKS system, and discuss the numerous ways in which the use of Semantic Web concepts and DAML extend its ability to provide an intelligent online service to both the human community and the agents assisting them.

1 Introduction

With the vast quantity of information now available on the Internet, there is a need to manage this information by marking it up with a semantic language, such as DARPA Agent Markup Language (DAML) [26], and using intelligent search engines, in conjunction with ontology-based matching, to provide more efficient and accurate information search results. The aim of the Semantic Web is to make the present web more machine-readable, in order to allow intelligent agents to retrieve and manipulate pertinent information. The key goal of the DAML program is to develop a Semantic Web markup language that provides sufficient rules for ontology development [20] and that is sufficiently rich to support intelligent agents and other applications [22,32]. Today's agents are not tightly integrated into the web infrastructure. If our goal is to have agents acting upon and conversing about web objects, they will have to be seamlessly integrated with the web, and take advantage of existing infrastructure whenever possible (e.g., message sending, security, authentication, directory services, and application service frameworks). We believe that DAML will be central to the realization of this goal.

* This work was supported in part by the Defense Advanced Research Projects Agency under contract F30602-00-2-0 591 AO K528 as part of the DAML program (`http://daml.org/`).

A. Omicini, P. Petta, and R. Tolksdorf (Eds.): ESAW 2001, LNAI 2203, pp. 175–193, 2001.

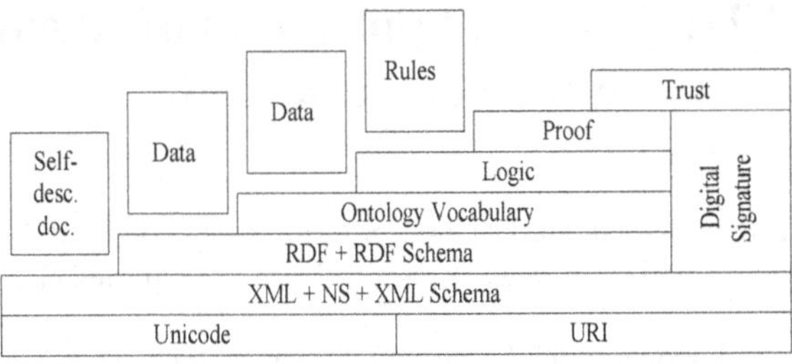

Fig. 1. Tim Berners-Lee's vision of the Semantic Web is founded on a base that includes URIs, XML, and RDF.

In support of this claim, we have constructed a real, fielded application, ITTALKS, which supports user and agent interaction in the domain of talk discovery. It also provides a simple web-driven infrastructure for agent interaction. In addition, ITTALKS serves as a platform for designing and prototyping the software components required to enable developers to create intelligent software agents capable of understanding and processing information and knowledge encoded in DAML and other semantically rich markup languages. To date, we have focused on developing the support and infrastructure required for intelligent agents to integrate into an environment of web browsers, servers, application server platforms, and associated supporting languages (e.g., WEB/SQL, WEBL), protocols (e.g., SSL, S/MIME, WAP, eSpeak), services (e.g., LDAP) and underlying technologies (e.g., Java, Jini, PKI).

On the surface, ITTALKS is a web portal offering access to information about talks, seminars and colloquia related to information technology (IT). It is organized around domains, which typically represent event hosting organizations such as universities, research laboratories or professional groups, and which in turn are represented by independent web sites. ITTALKS utilizes DAML for its knowledge base representation, reasoning, and agent communication. DAML is used to markup all the information, which is stored in a knowledge base, to provide additional reasoning capabilities otherwise unavailable. With information denoted in a semantically machine-understandable format, the computer can deduce additional information, a task which is difficult in a traditional database system. For example, if both ITTALKS and the user agree on a common semantics, the ITTALKS web portal can provide not only the talks that correspond to the user's profile in terms of interest, time, and location constraints, but can further filter the IT events based on information about the user's personal schedule, inferred location at the time of the talk, distance and current traffic patterns, etc. ITTALKS can also dynamically update the user's profile with incremental learning of the user's usage patterns.

ITTALKS demonstrates the power of markup languages such as DAML for the Semantic Web, drawing on its ability to represent ontologies, agent content languages and its ability to improve the functionality of agents on the web. We have developed DAML-encoded ontologies for describing event, temporal, spatial, personal, and conversational information, which enable us to represent all required knowledge in a DAML-encoded format. Moreover, these ontologies enable us to execute a computer understandable conversation. In addition, we have created several DAML-encoded classification ontologies, which provide us with additional reasoning capabilities in order to find the best matching IT talks for a particular user. Furthermore, in the ITTALKS application, any web page presented on the ITTALKS web sites contains the necessary information for an agent to retrieve the DAML-encoded description of this page as well as the contact information of a responsible agent in order to provide more effective conversation. ITTALKS thus provides each agent with the capability to retrieve and manipulate any ITTALKS-related information via a web site interface or through a direct agent-to-agent conversation. Hence, by combining the features of currently existing web applications with the DAML-based knowledge and reasoning capabilities, ITTALKS presents a true Semantic Web application.

2 Background

DAML is a semantic language being developed by a consortium of U.S.-based academic and business researchers [1], which officially began in August 2000, to address the current web's limitations in providing machine-readable, and more importantly machine-interpretable, information over the Internet. The goal of DAML is to enable the transformation of the currently human-oriented web, which is largely used as a text and multimedia repository only, into a Semantic Web as envisioned by Berners-Lee [4,3]. This process involves the augmentation of web pages with additional information and data that are expressed in a way that facilitates machine understanding [21,23].

DAML is built upon the capabilities of an already existing syntactic language, the Extendable Markup Language (XML) [19], and of the Resource Description Framework and Resource Description Framework Schema (RDF/S) [16,18,17,36]. These are XML applications that provide a number of preliminary semantic facilities required in the realization of the Semantic Web vision.

XML was developed by the World Wide Web Consortium (W3C) as a standard for alternative data encoding on the Internet that was primarily intended for machine processing. Moreover, XML is an application profile of the Standard Generalized Markup

[1] Visit the DAML Program official web site at `http://www.daml.org/`.

```
<daml:class rdf:ID="Animal">
 <rdfs:label>Animal</rdfs:label>
 <rdfs:comment>An Example</rdfs:comment>
</daml:Class>

<daml:Class rdf:ID="Male">
 <rdfs:subClassOf rdf:resource="#Animal"/>
</daml:Class>

<daml:class rdf:ID="Female">
  <rdfs:subClassOf rdf:resource="#Animal" />
  <daml:disjointWith rdf:resource="#Male" />
</daml:Class>

<daml:Class rdf:ID="Man">
  <rdfs:subClassOf rdf:resource="#Person" />
  <rdfs:subClassOf rdf:resource="#Male" />
</daml:Class>
```

Fig. 2. An Example of DAML-Encoded Knowledge.

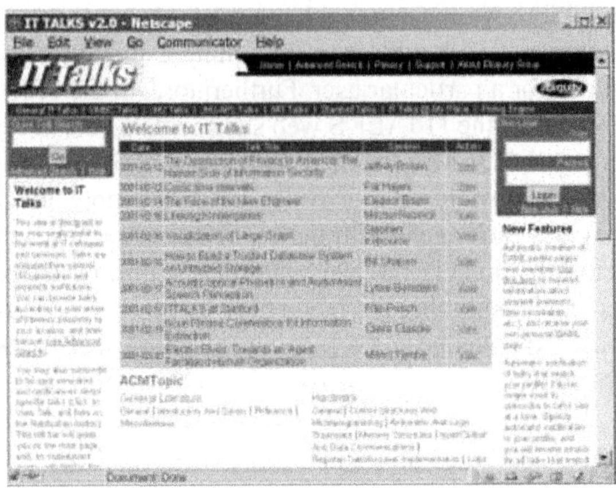

Fig. 3. A Screenshot Depicting the Main Page of the ITTALKS System.

Language (SGML), and is therefore based on a well-defined and well-understood syntactic language. The XML standard provides the necessary means to declare and use simple data structures, which are stored in XML documents and which are machine-readable. Subsequently, the information made available in these documents can be processed or translated into additional XML documents to provide the appropriate form for human understanding, such as text-to-voice, graphics or HTML conversion. However, since XML is defined only at the syntactic level, machines cannot be relied upon to unambiguously determine the correct meaning of the XML tags used in a given XML document. Consequently, XML is not suitable as a desired language for representing complex knowledge.

As a result, the W3C Consortium has developed RDF/S with the goal of addressing the XML deficiencies by adding formal semantics on the top of XML. These two standards provide the representation frameworks for describing relationships among resources in terms of named properties and values, which are similar to representation frameworks of semantic networks and rudimentary frame languages such as RDF Schema. Yet, both standards are still very restricted as a knowledge representation language due to the lack of support for variables, general quantification, rules, etc.

DAML is an attempt to build upon XML and RDF/S to produce a language that is well suited for building the Semantic Web. It follows the same path for representing data and information in a document as XML, and provides similar rules and definitions to RDF/S. In addition, DAML also provides rules for describing further constraints and relationships among resources, including cardinality, domain and range restrictions, as well as union, disjunction, inverse and transitive rules. DAML is, therefore, an effort to develop a universal Semantic Web markup language that is sufficiently rich to provide machines not only with the capability to read data but also with the capability to interpret and infer over the data. DAML will enable the development of intelligent agents and applications that can autonomously retrieve and manipulate information on the Internet and from the Semantic Web of tomorrow.

3 ITTALKS

As part of UMBC's role in the DAML Program, we have developed ITTALKS; a web portal that offers access to information about talks, seminars, colloquia, and other information technology (IT) related events. ITTALKS provides users with numerous details describing the IT events, including location, speaker, hosting organization, and talk topic. More importantly, ITTALKS also provides agents with the ability to retrieve and manipulate information stored in the ITTALKS knowledge base. Below, we discuss various aspects of the system in more detail.

Unlike other web services, ITTALKS employs DAML for knowledge base representation, reasoning, and agent communication. The use of DAML to represent information in its knowledge base, in conjunction with its use for interchangeable type ontologies as described in Section 5.6, enables more sophisticated reasoning than would otherwise be available. For example, a simpler representation scheme might be able to provide the user with talks based on interest, time and location. When both ITTALKS and the user agree on a common semantics, the ITTALKS web portal will be able to perform further filtering, based on more sophisticated inference. In addition to enhancing knowledge representation and reasoning, DAML is used for all communication, including simple messages and queries, using the ITTALKS defined ontology. Moreover, ITTALKS offers the capability for each user to use his/her personal agent to communicate with ITTALKS on his/her behalf and provide a higher level of service.

3.1 Users

ITTALKS can be used anonymously, or, more effectively, with personalized user accounts. Users have the option to register with ITTALKS either by entering information manually via web forms, or by providing the location (URL) of a universally accessible DAMLized personal profile, which includes information such as the users location, his/her interests and contact details, as well as a schedule. This schedule might be as rudimentary as a list of available time periods for given days, or could even include a detailed schedule for each day. Subsequently, this information is used to provide each user with a personalized view of the site, displaying only talks that match the user's interests and/or schedule.

Since DAML is not yet in widespread use, ITTALKS provides a tool for creating a DAML personal profile. Currently, the tool constructs a profile containing only items used by the ITTALKS system. However, we believe that the profile, in one form or another, will ultimately provide a unique and universal point for obtaining personal information about the user, not just for ITTALKS, but for all information needs, and will include any sort of information the user would like to share. In the future, all services that require personal information about the user should access the same user profile, eliminating the need for the user to repeatedly enter the same information for a multitude of services. We believe that the new standard for XML Signature and Encryption under development may provide a mechanism by which users can have some control over access to parts of their profile.

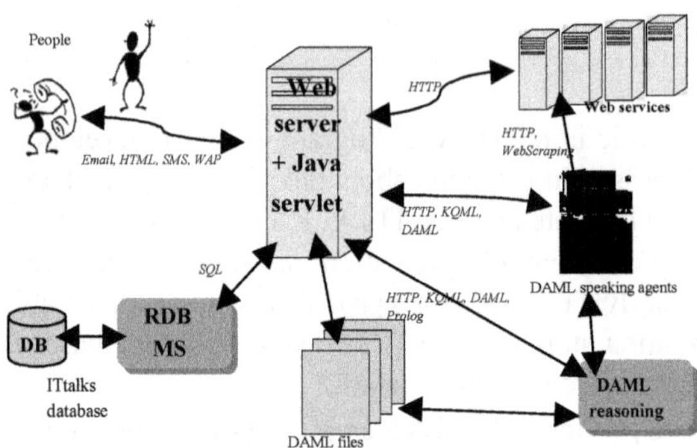

Fig. 4. The architecture for ITTALKS is built around a web server backed by a relational database. Interfaces are provided to human users, software agents and web services.

3.2 Domains

To support our vision of a universal resource for the international IT research community, ITTALKS is organized around domains, which typically represent event hosting organization such as universities, research laboratories or professional groups. Each domain is represented by a separate web site and is independently maintained by a moderator who can, among other things, define the scope of the domain and delegate to other registered users the ability to edit talk entries. For example, the `stanford.ittalks.org` domain might be configured to include only talks hosted at Stanford University. On the other hand, another domain, `sri.ittalks.org`, might be configured to include not only talks about Semantic Web topics that are held at SRI, but also those at Stanford, as well as any talks within 15 mile range of the SRI facility in Palo Alto.

3.3 Access

The ITTALKS system is accessible either to users directly via the web, or to agents acting on their behalf. The web portal provides numerous features, including registration, search, entry and domain administration. An agent-based interface allows interaction with user agents or other services.

Human Interface. The web portal allows a user to browse desired information in a variety of formats, to provide the highest degree of interoperability. It permits a user to retrieve information in DAML, standard HTML format, which includes a short DAML annotation for DAML-enabled web crawlers, or WML [11] format, which supports WAP enabled phones. The ITTALKS web portal also has the ability to generate RDF Site Summary (RSS) [34] files for certain queries. These RSS files can then be used for various external purposes, such as displaying upcoming talks on a departmental web site for some particular university or domain.

Agent Interface. To provide access for agent based services, ITTALKS makes use of Jackal [12], a communication infrastructure for Java-based agents developed by our research group at UMBC. Jackal is a Java package, which provides a comprehensive communications infrastructure while maintaining maximum flexibility and ease of integration. The heart of Jackal is a simple conversation system, serving to maintain context for concurrent threads of conversation while providing a guide for judging behavioral correctness and modeling the actions of other agents. Jackal provides facilities for creating and manipulating user-defined conversation structures of arbitrary extent. Jackal has a very modular, loosely coupled architecture, designed to support maximal concurrency among components, accomplished with the use of multiple threads and buffered interfaces between subsystems. Its concise API allows for comprehensive specification of message requests, and for blocking or non-blocking message retrieval. Currently, it facilitates the use of KQML agent communication language [14] and employs a sophisticated protocol for agent naming, addressing and identity (KNS). Additionaly, it is in the process of adapting to the FIPA standards [15,2]. In addition, our research group, in cooperation with other universities, is developing a DAML ontology for the necessary conversation protocols.

3.4 Agents

In order to extend the capabilities of the system, we have defined a number of agents that support the operation of ITTALKS. Some can be seen as supporting services (such as external information services), while others we assume will exist in the general environment in the future.

ITTALKS Agent. The ITTALKS agent is a front-end for the ITTALKS system. It interacts with ITTALKS through the same web-based interface as human users, but communicates via an ACL with other agents on the web, extending the system's accessibility. At present, the agent does not support any advanced functionality, but acts primarily as a gateway for agent access.

User Agents. One longtime goal of agent research is that users will be represented online by agents that can service queries and filter information for them. While ITTALKS does not require that such agents exist, we recognize the added power that could be gained by the use of such agents. Therefore, ITTALKS supports interaction with User Agents as well as their human counterparts. The User Agent that we have developed understands DAML, supports sophisticated reasoning, and communicates via a standard agent communication language. Reasoning is accomplished with the use of the XSB, a logic programming and deductive database system for Unix and Windows developed at SUNY Stony Brook.

Calendar Agent. Although a user agent may contain the necessary knowledge about its user's schedule, we believe that it will benefit from assigning the calendar-based facts and preferences to a separate agent - the calendar agent. This enables the user

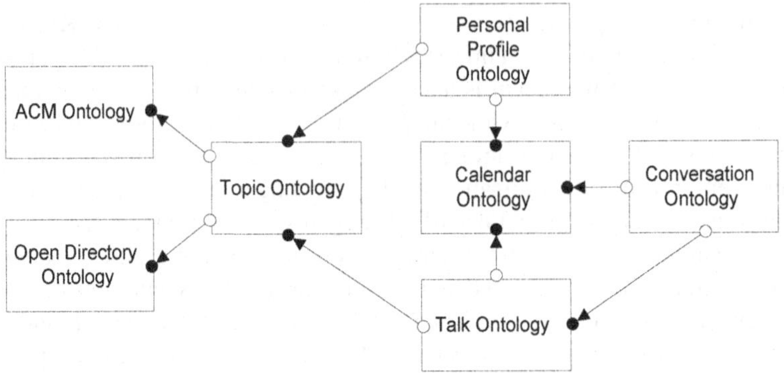

Fig. 5. The Relationships among the Various Ontologies Used by the ITTALKS System.

agent not only to consult the user calendar, but also to use the same protocol to consult other calendar agents that may represent other users or groups the user belongs to. In addition, the calendar agent may only represent abstraction to already existing infrastructure, such as Microsoft Outlook [33] or other desktop/server applications. Finally, the calendar agent may also be used to represent a room, and thus allow for re-use of the same principles of participation scheduling as well as event scheduling.

Classifier Agent. ITTALKS uses a Classifier (or recommender) Agent that is invoked when a user is entering a new talk. Based on the talk's abstract, the Classifier returns ACM Classification Hierarchy Classification numbers along with a rank, in descending order. Using a local table of classification numbers and names, ITTALKS suggests to the user ten possible topics.

MapQuest Agent. The MapQuest Agent is a wrapper agent that allows ITTALKS to make use of external services. It interacts directly with agents (e.g. the ITTALKS agent, User Agents), and accepts requests for information such as the distance between two known locations. It then phrases an appropriate request to the MapQuest system [29], parses the results, and generates an appropriate response. Note that this agent could be generically named a Distance Agent, and make use of any external service (or combination of several, as needed).

3.5 Ontologies

The ITTALKS system is based on a set of ontologies[2] that are used to describe talks and the things associated with them, e.g., people, places, topics and interests, schedules, etc. Figure 5 shows some of the dependencies that exist among these ontologies. The ontologies are used in the representation and processing of DAML descriptions and

[2] `http://daml.umbc.edu/ontologies.`

also as "conceptual schemata" against which the database and various software APIs are built.

We have developed a general ontology for describing the topics of arbitrary talks and papers. Using this, we have implemented an ontology to describe IT related talks based on the ACM's Computer Classification System. In addition, we currently are developing a DAML ontology for IT talks based on a portion of the Open Directory, and are considering additional classification ontologies. Figure 7 sketches some of the major classes and properties in these ontologies. These topic ontologies are used to describe talks as well as the users' interests throughout the system. This includes an automated talk classification, for which we have obtained a training collection for the ACM CCS and are also generating an Open Directory training collection to develop the necessary components. In addition, the DAML ontologies will give a user the ability to add additional assertions in DAML to further characterize their interests. Lastly, we are also in the process of developing a semi-automated component that can map topics in one ontology into topics in another, by utilizing user-specific mapping believes and by taking advantage of the fact that nodes in each ontology have an associated collection of text. This component is further described in Section 5.4.

3.6 Data Entry

Currently ITTALKS requires that information about talks be manually entered via a web form interface, or be available in a DAML description available at a given URL. Although we have made this process as simple as possible (e.g., by supporting automatic form completion using information from the knowledge base and the user's DAML profile) it is still a time consuming process. Therefore, we are developing a focused web spider to collect talk announcements from open sources on the web. This spider will identify key information items using a text extraction system, and will automatically add information to the ITTALKS knowledge base. We are working with the Lockheed-Martin research group on the above task, and will use their AeroText information extraction system [1].

3.7 Architecture

The current implementation of ITTALKS uses a relational database, in combination with a web server, to provide user access to the system. To enable agents to access the system, the ITTALKS provides an interface for agent-based communication.

Database. The main software packages that are used in the ITTALKS system are the MySQL relational database software and a combination of Apache and Tomcat as the web portal servers. The contents of the ITTALKS knowledge base are stored in a database whose schema is closely mapped to our ontologies describing events, people, topics and locations. We have chosen MySQL because of its known reliability, and because we required software with a license that allows us to make the ITTALKS package available to additional academic and commercial institutions.

Web Server. As stated above, for our web, we have chosen a combination of Apache and Tomcat. This enables us to present the IT talk descriptions to the user using Java servlets and JSP files, which dynamically generate requested information in DAML, XML, HTML, RSS, and WML formats. The current ITTALKS implementation can provide information suitable for viewing on either a standard, computer-based or a WAP-enabled cellular phone.

Extensions. In addition, we are currently employing the Jackal agent communication infrastructure developed at UMBC and the Lockheed-Martin's AeroText information extraction system in order to facilitate ITTALKS-user agent interaction and the automated text extraction, respectively. We are in the process of modifying Jackal to provide support for FIPA ACL interoperability. Also, we are considering the possible replacement of MySQL with native XML database software such as dbXML.

4 Scenarios

We describe here a couple of typical interactions that illustrate some of the features of ITTALKS. The first involves direct use by a human user, and the second, advanced features provided through the use of agents.

4.1 Human Interaction

In this first scenario, a user, Jim, learns from his colleagues about the existence of the ITTALKS web portal as a source of IT related events in his area; Jim is affiliated with Stanford University.

Jim directs his browser to the `http://www.ittalks.org` main page. Seeing a link to `http://stanford.ittalks.org` (a Stanford ITTALKS domain), he selects it, and is presented with a new page listing upcoming talks that are scheduled at Stanford, SRI and other locations within a 15-mile radius (the default distance for the Stanford domain).

Jim browses the web site, viewing announcements for various talks matching his interests and preferred locations (as provided in his explicit search queries). He is impressed that he can see the talk information not only in HTML, but also in DAML, RSS and WML formats. Finding a talk of potential interest to a colleague, Jim takes advantage of the invitation feature, which allows him to send an invitational e-mail to any of his friends for any of the listed talks. Finally, using the personalize link on the bottom of the page, Jim creates his own ittalks.org main page, by providing the URL of his DAML-encoded profile. This customized page, listing talks based on his preferences, will be Jim's entrance to the ITTALKS site whenever her returns.

4.2 Agent Interaction

This scenario assumes that user Jim has already registered with ITTALKS, and has left instructions with the system to be notified of the occurrence of certain types of talks.

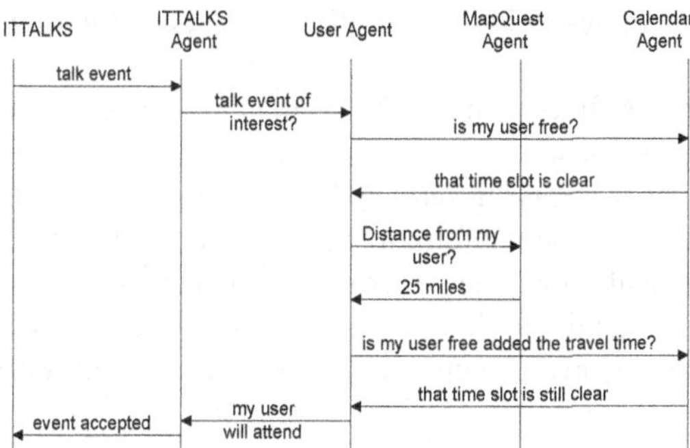

Fig. 6. Interactions between the Various Agents Described in the ITTALKS/Agent Scenario.

In the course of operation, ITTALKS discovers that there is an upcoming talk that may interest Jim, and of which Jim has not been notified. Based on information in Jim's preferences, which have been obtained from his online, DAML-encoded profile and from information entered directly, ITTALKS opts to notify Jim's User Agent directly. This is done via ITTALKS own agent, which forwards the message using an ACL.

Upon receiving this information, Jim's User Agent needs to know more; it consults with Jim's Calendar agent to determine his availability, and with the MapQuest agent to find the distance from Jim's predicted location at the time of the talk. Some more sophisticated interactions might take place at this time; for example, the Calendar and User agents may decide to alter Jim's schedule, and proceed to contact the User agent of some other individual. In addition, the User agent may request more information about the speaker and the event by contacting other agents or web sites, such as CiteSeer-based agent [5,30,6], to obtain more information necessary to make a decision. Finally, after making this decision, the User Agent will send a notification back to the ITTALKS agent indicating that Jim will/will not plan to attend. The ITTALKS agent will make the appropriate adjustments at the ITTALKS site.

In a more complex interaction scheme, Jim may be employed by a research group, which possesses a limited funding and is therefore enforcing a policy that allows only one researcher at a time to attend a particular IT event. As a result, the User agent cannot decide on Jim's participation until it successfully interacts with other agents representing Jim's employer and colleagues. Therefore, the decision whether anyone from the research group could attend the IT event and the further election of the group representative requires an interaction of agent virtual community.

From a slightly different perspective, the User agent will also benefit from participating in virtual communities thanks to recommendations it obtains from other User agents. One User agent may recommend an IT event given its owner's experiences from attending a past talk of the same speaker. Another User agent may decide to share comparisons of two competing times and locations for an identical IT event. Yet another User agent may simply share its owner's intensions on attending a particular IT event. Thus, each member of the virtual community can profit from these and many other positive and negative recommendations, and reflect these social filtering methods in its own decisions.

Finally, in a 'Smart Office' scenario [25,7], the ITTALKS agent may also be directly contacting an agent representing the location where a given IT event will be held. This 'room' agent may then use varying service discovery techniques [8,35] to locate a projector presented in the room and inform it to pre-download the powerpoint presentation before the speaker arrival. Moreover, the 'room' agent may also try to contact additional agents in the IT event vicinity to decrease possible noise level from other rooms and to verify that a 'hallway' agent has requested enough refreshments during the IT event.

5 Benefits of DAML

We believe that ITTALKS benefits significantly from its use of a semantic markup language such as DAML. DAML is used to specify ontologies that we use extensively in our system. It is also used for personal profiles, and as an agent content language. Without DAML, specifying schedules, interests and assertions about topics would be very difficult. In ITTALKS, a user can specify that according to the user a couple of topics are equivalent or related or dissimilar, etc. This will allow ITTALKS to tailor the searching of talks to the users needs. As an agent content language, DAML provides more flexible semantics than KIF or other content languages that currently provide syntax only. The ultimate benefit of using DAML then lies in the ability of ITTALKS to independently interact with any DAML-capable agent without the need of a human supervision. Consequently, all these benefits, which are described in further details below, enable more efficient interaction between the system and its users, let them be humans or software agents.

5.1 Interoperability Standard

As an interoperability layer, DAML allows the content of ITTALKS to be easily shared with other applications and agents. For example, a Centaurus room manager agent [25] could watch ITTALKS for events happening in a room for which it is responsible in order to enable better scheduling. DAML also acts as an interoperability standard allowing other sites to make their talks available for inclusion in ITTALKS by publishing announcements marked up in our ontology.

5.2 Agent Communication Language

DAML and ACLs can be successfully integrated. DAML documents will be the objects of discourse for agents that will create, access, modify, enrich and manage DAML documents as a way to disseminate and share knowledge. Agents will need to communicate with one another not only to exchange DAML documents but also to exchange *informational attitudes* about DAML documents. Using an Agent Communication Languages (ACL) agents can "talk" about DAML documents. Integrating ACL work and concepts with a universe of DAML content is our first goal. Using DAML as an agent content language will add more meaning to the message.

5.3 Distributed Trust and Belief

Agents face a difficult problem of knowing what information sources (e.g. documents, web pages, agents) to believe and trust in an open, distributed and dynamic world, and how to integrate and fuse potentially contradictory information. DAML can be used to support *distributed trust and reputation management* [24,27,28]. This will form the basis of a logic for *distributed belief transfer* that will enable more sophisticated, semantically-driven rule-based techniques for information integration and fusion.

We are making use of DAML's expressiveness and employing it to describe security policies, credentials and trust relationships, which form the basis of trust management. These policies contain more semantic meaning, allowing different policies to be integrated and conflicts to be resolved relatively easily. Also, it will be possible for other applications to interpret the agent's credentials, e.g. authorization certificates, correctly, making these credentials universal.

Similarly, describing beliefs and associating levels of trust with these beliefs is more straightforward and the deduction of belief is uniform by different applications and services.

5.4 Data Entry Support

ITTALKS supports intelligent form filling, making it easier for users to enter and edit information in their profiles, and also to enter and edit talk announcements and other basic information. In addition, we provide automatic form filling when an editor tries to enter information about an entity (e.g. a talk, person, room) that already present in the knowledge base.

Entering Talks. In order to make ITTALKS successful, as we need to make it as easy as possible new talk descriptions to be entered into the system. We are addressing this problem using three complimentary approaches: an enhanced web interface, accepting marked up announcements, and automated text extraction. DAML plays a key role in the first two and is the target representation for the third.

Enhancing the Web Interface. We have used several techniques to enhance the web form interface for entering talk announcements. One of the simplest and most effective is to recognize then some of the information being entered about an object such as a person, a room or an organization has already been entered into the ITTALKS system and to "pre-fill" the remaining parts of the form from our stored information. For example, most talks at an organization are given in a small number of rooms. Once the complete information about a particular room (e.g., room number, building, address, seating capacity, longitude and latitude, A/V equipment, networking connection, etc.) has been entered for one talk, it need not be entered again.

Although the current implementation of this does not directly use DAML, its use can support a more generalized version of a web form-filling assistant. The approach depends on two ideas: (i) tagging web form widgets with DAML descriptions of what they represent and (ii) capturing dependencies among data items in DAML and (iii)

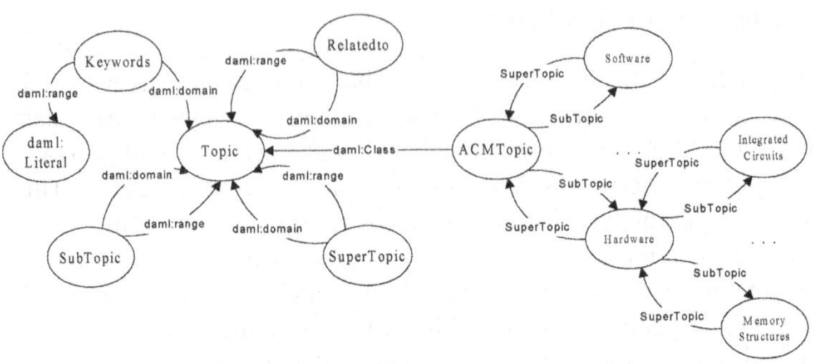

Fig. 7. The Ontologies used by IITALKS are relatively simple, such as the topics ontology used to describe talk topics and user interests.

compiling these dependencies into an appropriate execution form (e.g., JavaScript procedures) that can drive the web form interface.

In addition, we plan to investigate the possibility of a multi-modal support, where user can enter new information via standard keyboard input as well as through voice recognition means. Here, we understand that when presenting the user with a new form, the user will be allowed to use her own voice to enter data in each field. Then upon submittion of voice-filled form, ITTALKS will try to infer the meaning of the recorded sound, obtain additional information based on the knowledge and rules stored in ITTALKS system, and present back the user with a text-prefilled form for verification purposes. This enhancement will then allow ITTALKS to provide talk entry support for devices with limited keyboard functionality, such as PDAs or cellular phones.

Text Classification. For ITTALKS to filter talk announcements on topic matches, we need to know the appropriate topics for each talk. Initially, we required that users manually select appropriate topic categories from a web interface to the ACM CCS hierarchy. This turns out to be a daunting task requiring the user to navigate in a hierarchy of nearly 300 topics, many of which about whose meaning he will not be sure. Some users will face a similar problem in trying to select topics to characterize their own interests. Ultimately we would like to use more that one topic hierarchy to classify both talk topics and user interests (e.g., ACM CCS and Open Directory nodes), which makes the problem even more difficult for our users.

To address this problem, we have built an automatic text classifier that can suggest terms in a hierarchy that are appropriate for classifying a talk based on its title and abstract. The classifier package used was from the Bag Of Words (BOW) toolkit [31] by Andrew McCallum at CMU. This library provides support for a wide variety of text classification and retrieval algorithms. We used the Naive Bayes algorithm, which is widely used in the classification literature, fairly effective, and quick to learn the 285 classes in our test collection. We plan to use the same classification agent to suggest interest terms for users based on the text found by searching their web pages.

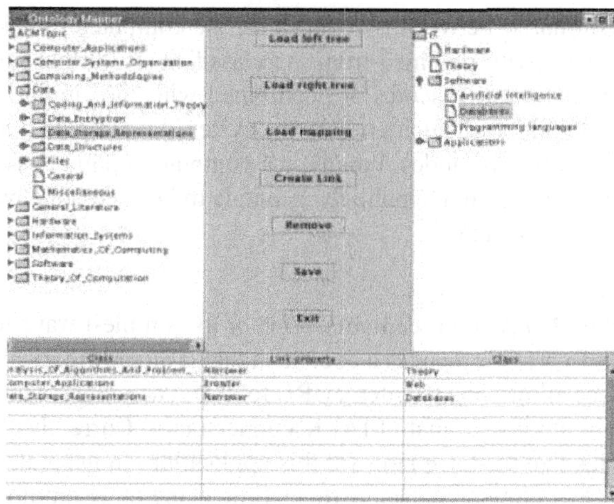

Fig. 8. Snapshot of the Ontology Semi-Automatic Mapper Prototype.

In addition to classifying text into a particular ontology at a time, we are also developing a tool for mapping among multiple ontologies. Such tool will, for example, allow each user to select her preferred topic ontology on the fly, and the ITTALKS system will then immediately adapt and present the personalized filtering results according to this ontology. As a prototype of the mapping tool, we have chosen a semi-automatic approach, wherein user can manually select relations among specific topics accross the ontologies ("landmarks"), e.g. broader, narrower, and similar. Subsequently, the remaining mappings are automatically computed via a combination of the user-specified relations and our automatic text classifier using training sets of documents for the ontologies. The automated mapper then operates in the following manner:

Consider two topic ontologies A and B, each of which is a strict hierarchy. Each node in each hierarchy $(A_1, A_2, ..., A_n)$, $(B_1, B_2, ..., B_m)$ has a set of exemplary documents that have already been classified as being associated with that node. Let $parent_i$ be the parent of node i and $text_i$ be the string of all the text associated with node i. We have used a text classifier to build a classifier for ontology A and another one for ontology B. The classifier for ontology A is built taking an arbitrary string S to be classified and computing the function $S_a(S, A_i)$ for each node A_i in ontology A. The classifier for ontology B computes $S_b(S, B_i)$. These similarity scores could further be normalized as real numbers between 0 and 1. These two classifiers are used to compute two raw topic similarity matrices $SM_{ab}(A_i, B_j)$ and $SM_{ba}(A_i, b_j)$, for each pair of nodes, one from ontology A and one from ontology B. Therefore,

$$SM_{ab}(A_i, B_i) = S_b(text(A_i), B_j)$$

$$SM_{ba}(A_i, B_j) = S_a(text(B_j), A_i)$$

One simple way to compute the similarity of topic i in ontology A to topic j in ontology B would be to obtain a weighted sum of these two measures:

$$S_{ab}(A_i, B_j) = W * SM_{ab}(A_i, B_j) + (1 - W) * SM_{ba}(A_i, B_j)$$

In this way, the similarity between two nodes can be computed by taking a weighted sum of their raw similarity obtained from the classifier. To further improve mapping results, the the raw similarity scores are combined with the similarity scores of the nodes' parents and children, which enables the system to compute a more precise relations between each pair of nodes. Finally, the computed relations are converted into a DAML file, which is then made publicly available to be accessible by ITTALKS and other Semantic Web applications.

Accepting Marked Up Announcements. One of the simplest ways to enter new talk announcements is to provide them as a document that is already marked up. The current ITTALKS interface allows one to enter a URL for a talk announcement that is assumed to be marked up in ontologies that ITTALKS understands. Currently, these are just the "native" ontologies that we have built for this application. In general, if some talk announcements were available with semantic markup using other ontologies, it might be possible to provide rules and transformation that could map or partially map the information into the ITTALKS ontologies. We expect that, as the Semantic Web develops, it will be more and more likely that talk announcements with some meaningful mark up will be found on the web.

Automated Information Extraction from Text. We would like to be able to process talk announcements in plain text or HTML and automatically identify and extract the key information required by ITTALKS. This would allow us to fill the ITTALKS database with information obtained from announcements delivered via email lists or found on the web. The problem of recognizing and extracting information from talk announcements has been studied before [13,10] mostly in the context of using it as a machine learning application. We are developing a information extraction use the Aerotext [1,9] system that can identify and extract the information found in a typical talk announcement and use this to automatically produce a version marked up in DAML which can then be entered in the ITTALKS database.

5.5 User Profiles

We use personal profiles to help ITTALKS meet the requirements of individual users. A profile is a widely accessible source of information about the user, marked DAML, to which other services and individuals can refer. In the future, such a profile may be used by all web-based services that the user wants to access. The profile will ultimately provide a unique and universal point for obtaining personal information about the user for all services, preventing the need for duplication and potential inconsistencies. This profile can be easily shared, and with the use of DAML, will allow more expressive content for schedules, preferences and interests. The notion of a personal profile and a user agent are closely linked; a user might have one or the other, or both. The profile would likely express much of the information that might be encoded in a user agent's knowledge base. Conversely, an agent would likely be able to answer queries about information contained in a profile.

5.6 Modularity

With the use of DAML, we can define several ontologies for topics and switch between them with ease. Furthermore, to restrict the retrieval results, a user can perform the search with respect to a certain set of ontologies, such as the ACM or Open Directory Classification.

5.7 Application Scalability Support

As ITTALKS becomes the central repository of IT related information for various research institutes the ITTALKS knowledge base will be distributed among numerous, and possibly apriori-unknown, locations in order to provide a higher scalability and reliability support. Yet, it will be imperative that users and agents not be required to interact with all locations in order to find or manipulate the desired information. Instead, we envision that each user agent will interact with only one ITTALKS agent, which in turn will be able to efficiently locate and manage the distributed ITTALKS information. For this, we believe that a system of DAML-enabled agents can act as an intermediate between the distributed databases.

6 Future Directions

Since most users do not currently have personal agents, we have been developing one that can be used with this system. It is our goal, however, that ITTALKS be able to interact with external agents of any type. The agent we are developing reasons about the user's interests, schedules, assertions and uses the MapQuest agent to figure out if a user will be able to attend an interesting talk on a certain date.

We are developing a framework to use DAML in distributed trust and belief. DAML expressions on a web page that encodes a statement or other speech act by an agent are signed to provide authentication and integrity. We are working on an ontology to describe permissions, obligations and policies in DAML and allow agents to make statements about and delegate them.

In order to make the process of data entry more efficient, we are developing a focused web spider, which will collect talk announcements from source on the web and to identify the key information in these announcements using a text extraction system. The spider will add all found and relevant information to the ITTALKS knowledge base.

7 Conclusion

Effective use of the vast quantity of information now available on the web necessitates semantic markup such as DAML. With the use of such a tool, we can enable the automated or machine-facilitated gathering and processing of much information that is currently lost to us. ITTALKS, our system for automatic and intelligent notification of Information Technology talks, demonstrates the value of DAML in a variety of ways. DAML is used throughout the ITTALKS system, from basic knowledge representation, to inter-agent communication.

References

1. AeroText. site:
`http://mds.external.lmco.com/products_services/aero/`.
2. Fabio Bellifemine, Agostino Poggi, and Giovanni Rimassa. Developing multi agent systems with a fipa-compliant agent framework. *Software - Practice and Experience*, 3, 2001.
3. Tim Berners-Lee and Mark Fischetti. Weaving the web: The original design and ultimate destiny of the world wide web by its inventor. *Harper, San Francisco*, 2001.
4. Tim Berners-Lee, James Hendler, and Ora Lassila. The semantic web. *Scientific American*, May 2001.
5. Kurt D. Bollacker, Steve Lawrence, and C. Lee Giles. *Citeseer: An autonomous web agent for automatic retrieval and identification of interesting publications*. Proceedings of the Second International Conference on Autonomous Agents (Agents'98). ACM Press, Minneapolis, 1998.
6. Sergey Brin and Lawrence Page. *The Anatomy of a Large-Scale Hypertextual Web Search Engine*. Proceedings of the 7th International World Wide Web Conference. April 1998.
7. Andrej Cedilnik, Lalana Kagal, Filip Perich, Jeffrey Undercoffer, and Anupam Joshi. A secure infrastructure for service discovery and access in pervasive computing. *Technical report, TR-CS-01-12, CSEE, University of Maryland Baltimore County*, 2001.
8. Dipanjan Chakraborty, Filip Perich, Sasikanth Avancha, and Anupam Joshi. Dreggie: Semantic service discovery for m-commerce applications. *Workshop on Reliable and Secure Applications in Mobile Environment, 20th Symposiom on Reliable Distributed Systems*, October 2001.
9. Lois C. Childs. Aerotext - a customizable information extraction system. unpublished technical report, Lockheed Martin, 2001.
10. Fabio Ciravegna. Learning to tag for information extraction from text. *ECAI Workshop on Machine Learning for Information Extraction*, August 2000. workshop held in conjunction with ECAI2000, Berline.
11. WAPForum Consorscium. Wireless markup language, November 1999. site:
`http://www1.wapforum.org/tech/documents/SPEC-WML-19991104.pdf`.
12. R. Scott Cost, Tim Finin, Yannis Labrou, Xiaocheng Luan, Yun Peng, Ian Soboroff, James Mayfield, and Akram Boughannam. Jackal: A java-based tool for agent development. *Working Notes of the Workshop on Tools for Developing Agents, AAAI'98*, July 1998.
13. T. Elliassi-Rad and J.Shavlik. Instructable and adaptive web-agents that learn to retrieve and extract information. *Department of Computer Sciences, University of Wisconsin, Machine Learning Research Group Working Pap*, 2000.
14. Tim Finin, Yannis Labrou, and James Mayfield. *Software Agents*, chapter KQML as an agent communication. MIT Press, Cambridge, 1997.
15. FIPA. Fipa 97 specification part 2: Agent communication language. *Technical report, FIPA - Foundation for Intelligent Physical Agents*, October 1997.
16. W3C Working Group. W3c resource description framework (rdf), October 1998. site: `http://www.w3c.org/RDF`.
17. W3C Working Group. W3c resource description framework model and syntax specification, February 1999.
18. W3C Working Group. W3c resource description framework schema (rdfs), March 1999. site: `http://www.w3.org/TR/rdf-schema/`.
19. W3C Working Group. extensible markup language (xml), October 2000. site: `http://www.w3.org/XML`.
20. N. Guarino. *Formal Ontology in Information Systems*, chapter Formal ontology and information systems. IOS Press, 1998.

21. Jeff Heflin, James Hendler, and Sean Luke. Shoe: A prototype language for the semantic web. *Linköping Electronic Articles in Computer and Information Science, ISSN 1401-9841*, 6, 2001.
22. James Hendler. Agents and the semantic web. *IEEE Intelligent Systems*, 16(2):30–37, March/April 2001.
23. James Hendler and Deborah McGuinness. The darpa agent markup language. *IEEE Intelligent Systems*, 15(6):72–73, November/December 2000.
24. Lalana Kagal, Harry Chen, Scott Cost, Timothy Finin, and Yun Peng. An infrastructure for distributed trust management. *Autonomous Agents Workshop on Norms and Institutions in Multiagent Systems, AA 2001, Montreal, Canada*, May 2001.
25. Lalana Kagal, Vlad Korolev, Harry Chen, Anupam Joshi, and Timothy Finin. A framework for intelligent services in a mobile environment. *Proceedings of the International Workshop on Smart Appliances and Wearable Computing (IWSAWC)*, April 2001.
26. DARPA Agent Markup Language. site: `http://www.daml.org/`.
27. Ninghui Li, Joan Feigenbaum, and Benjamin Grosof. A logic-based knowledge representation for authorization with delegation (extended abstract). *Proc. 12th IEEE Computer Security Foundations Workshop, Mordano, Italy*, June 1999. IBM Research Report RC 21492.
28. Ninghui Li and BBenjamin Grosof. A practically implementable and tractable delegation logic. *IEEE Symposium on Security and Privacy*, May 2000.
29. MapQuest. site: `http://www.mapquest.com/`.
30. James Mayfield, Paul McNamee, and Christine Piatko. The jhu/apl haircut system at trec-8. *The Eighth Text Retrieval Conference (TREC-8)*, pages 445–452, November 1999.
31. McCallum and Andrew Kachites. Bow: A toolkit for statistical language modeling, text retrieval, classification and clustering, 1996. site: `http://www.cs.cmu.edu/~mccallum/bow`.
32. Sheila A. McIlraith, Tran Cao Son, and Honglei Zeng. Semantic web services. *IEEE Intelligent Systems*, 16(2), March/April 2001.
33. Microsoft. Outlook. site: `http://www.microsoft.com/office/outlook/`.
34. Netscape. Rdf site summary (rss). `http://my.netscape.com/publish/formats/rss-spec-0.91`.
35. Olga Ratsimor, Vladimir Korolev, Anupam Joshi, and Timothy Finin. Agents2go: An infrastructure for location-dependent service discovery in the mobile electronic commerce environment. *ACM Mobile Commerce Workshop*, July 2001.
36. S. Staab, M. Erdmann, and A. Maedche. Ontologies in rdf(s). *Linköping Electronic Articles in Computer and Information Science*, 6, 2001. ISSN 1401-9841.

Author Index

Lecture Notes in Artificial Intelligence (LNAI)

Vol. 2005: W. Ziarko, Y. Yao (Eds.), Rough Sets and Current Trends in Computing. Proceedings, 2000. XV, 670 pages. 2001.

Vol. 2086: M. Luck, V. Mařík, O. Stěpánková, R. Trappl (Eds.), Multi-Agent Systems and Applications. Proceedings, 2001. X, 437 pages. 2001.

Vol. 2100: R. Küsters, Non-Standard Inferences in Description Logics. X, 250 pages. 2001.

Vol. 2101: S. Quaglini, P. Barahona, S. Andreassen (Eds.), Artificial Intelligence in Medicine. Proceedings, 2001. XIV, 469 pages. 2001.

Vol. 2103: M. Hannebauer, J. Wendler, E. Pagello (Eds.), Balancing Reactivity and Social Deliberation in Multi-Agent Systems. VIII, 237 pages. 2001.

Vol. 2109: M. Bauer, P.J. Gmytrasiewicz, J. Vassileva (Eds.), User Modeling 2001. Proceedings, 2001. XIII, 318 pages. 2001.

Vol. 2111: D. Helmbold, B. Williamson (Eds.), Computational Learning Theory. Proceedings, 2001. IX, 631 pages. 2001.

Vol. 2112: R. Kowalczyk, S.W. Loke, N.E. Reed, G. Williams (Eds.), Advances in Artificial Intelligence. Proceedings, 2000. IX, 363 pages. 2001.

Vol. 2116: V. Akman, P. Bouquet, R. Thomason, R.A. Young (Eds.), Modeling and Using Context. Proceedings, 2001. XII, 472 pages. 2001.

Vol. 2117: M. Beynon, C.L. Nehaniv, K. Dautenhahn (Eds.), Cognitive Technology: Instruments of Mind. Proceedings, 2001. XV, 522 pages. 2001.

Vol. 2120: H.S. Delugach, G. Stumme (Eds.), Conceptual Structures: Broadening the Base. Proceedings, 2001. X, 377 pages. 2001.

Vol. 2123: P. Perner (Ed.), Machine Learning and Data Mining in Pattern Recognition. Proceedings, 2001. XI, 363 pages. 2001.

Vol. 2132: S.-T. Yuan, M. Yokoo (Eds.), Intelligent Agents: Specification, Modeling, and Application. Proceedings, 2001. X, 237 pages. 2001.

Vol. 2143: S. Benferhat, P. Besnard (Eds.), Symbolic and Quantitative Approaches to Reasoning with Uncertainty. Proceedings, 2001. XIV, 818 pages. 2001.

Vol. 2148: A. Nareyek (Ed.), Local Search for Planning and Scheduling. Proceedings, 2000. IX, 171 pages. 2001.

Vol. 2155: H. Bunt, R.-J. Beun (Eds.), Cooperative Multimodal Communication. Proceedings, 1998. VIII, 251 pages. 2001.

Vol. 2157: C. Rouveirol, M. Sebag (Eds.), Inductive Logic Programming. Proceedings, 2001. X, 261 pages. 2001.

Vol. 2159: J. Kelemen, P. Sosík (Eds.), Advances in Artificial Life. Proceedings, 2001. XIX, 724 pages. 2001.

Vol. 2166: V. Matoušek, P. Mautner, R. Mouček, K. Taušer (Eds.), Text, Speech and Dialogue. Proceedings, 2001. XIII, 542 pages. 2001.

Vol. 2167: L. De Raedt, P. Flach (Eds.), Machine Learning: ECML 2001. Proceedings, 2001. XVII, 618 pages. 2001.

Vol. 2168: L. De Raedt, A. Siebes (Eds.), Principles of Data Mining and Knowledge Discovery. Proceedings, 2001. XVII, 510 pages. 2001.

Vol. 2173: T. Eiter, W. Faber, M. Truszczynski (Eds.), Logic Programming and Nonmonotonic Reasoning. Proceedings, 2001. XI, 444 pages. 2001.

Vol. 2174: F. Baader, G. Brewka, T. Eiter (Eds.), KI 2001: Advances in Artificial Intelligence. Proceedings, 2001. XIII, 471 pages. 2001.

Vol. 2175: F. Esposito (Ed.), AI*IA 2001: Advances in Artificial Intelligence. Proceedings, 2001. XII, 396 pages. 2001.

Vol. 2182: M. Klusch, F. Zambonelli (Eds.), Cooperative Information Agents V. Proceedings, 2001. XII, 288 pages. 2001.

Vol. 2190: A. de Antonio, R. Aylett, D. Ballin (Eds.), Intelligent Virtual Agents. Proceedings, 2001. VIII, 245 pages. 2001.

Vol. 2198: N. Zhong, Y. Yao, J. Liu, S. Ohsuga (Eds.), Web Intelligence: Research and Development. Proceedings, 2001. XVI, 615 pages. 2001.

Vol. 2203: A. Omicini, P. Petta, R. Tolksdorf (Eds.), Engineering Societies in the Agents World II. Proceedings, 2001. XI, 195 pages. 2001.

Vol. 2225: N. Abe, R. Khardon, T. Zeugmann (Eds.), Algorithmic Learning Theory. Proceedings, 2001. XI, 379 pages. 2001.

Vol. 2226: K.P. Jantke, A. Shinohara (Eds.), Discovery Science. Proceedings, 2001. XII, 494 pages. 2001.

Vol. 2246: R. Falcone, M. Singh, Y.-H. Tan (Eds.), Trust in Cyber-societies. VIII, 195 pages. 2001.

Vol. 2250: R. Nieuwenhuis, A. Voronkov (Eds.), Logic for Programming, Artificial Intelligence, and Reasoning. Proceedings, 2001. XV, 738 pages. 2001.

Vol. 2253: T. Terano, T. Nishida, A. Namatame, S. Tsumoto, Y. Ohsawa, T. Washio (Eds.), New Frontiers in Artificial Intelligence. Proceedings, 2001. XXVII, 553 pages. 2001.

Vol. 2256: M. Stumptner, D. Corbett, M. Brooks (Eds.), AI 2001: Advances in Artificial Intelligence. Proceedings, 2001. XII, 666 pages. 2001.

Vol. 2258: P. Brazdil, A. Jorge (Eds.), Progress in Artificial Intelligence. Proceedings, 2001. XII, 418 pages. 2001.

Lecture Notes in Computer Science

Vol. 2223: P. Eades, T. Takaoka (Eds.), Algorithms and Computation. Proceedings, 2001. XIV, 780 pages. 2001.

Vol. 2224: H.S. Kunii, S. Jajodia, A. Sølvberg (Eds.), Conceptual Modeling – ER 2001. Proceedings, 2001. XIX, 614 pages. 2001.

Vol. 2225: N. Abe, R. Khardon, T. Zeugmann (Eds.), Algorithmic Learning Theory. Proceedings, 2001. XI, 379 pages. 2001. (Subseries LNAI).

Vol. 2226: K.P. Jantke, A. Shinohara (Eds.), Discovery Science. Proceedings, 2001. XII, 494 pages. 2001. (Subseries LNAI).

Vol. 2227: S. Boztaş, I.E. Shparlinski (Eds.), Applied Algebra, Algebraic Algorithms and Error-Correcting Codes. Proceedings, 2001. XII, 398 pages. 2001.

Vol. 2228: B. Monien, V.K. Prasanna, S. Vajapeyam (Eds.), High Performance Computing – HiPC 2001. Proceedings, 2001. XVIII, 438 pages. 2001.

Vol. 2229: S. Qing, T. Okamoto, J. Zhou (Eds.), Information and Communications Security. Proceedings, 2001. XIV, 504 pages. 2001.

Vol. 2230: T. Katila, I.E. Magnin, P. Clarysse, J. Montagnat, J. Nenonen (Eds.), Functional Imaging and Modeling of the Heart. Proceedings, 2001. XI, 158 pages. 2001.

Vol. 2232: L. Fiege, G. Mühl, U. Wilhelm (Eds.), Electronic Commerce. Proceedings, 2001. X, 233 pages. 2001.

Vol. 2233: J. Crowcroft, M. Hofmann (Eds.), Networked Group Communication. Proceedings, 2001. X, 205 pages. 2001.

Vol. 2234: L. Pacholski, P. Ružička (Eds.), SOFSEM 2001: Theory and Practice of Informatics. Proceedings, 2001. XI, 347 pages. 2001.

Vol. 2235: C.S. Calude, G. Păun, G. Rozenberg, A. Salomaa (Eds.), Multiset Processing. VIII, 359 pages. 2001.

Vol. 2236: K. Drira, A. Martelli, T. Villemur (Eds.), Cooperative Environments for Distributed Systems Engineering. IX, 281 pages. 2001.

Vol. 2237: P. Codognet (Ed.), Logic Programming. Proceedings, 2001. XI, 365 pages. 2001.

Vol. 2239: T. Walsh (Ed.), Principles and Practice of Constraint Programming – CP 2001. Proceedings, 2001. XIV, 788 pages. 2001.

Vol. 2240: G.P. Picco (Ed.), Mobile Agents. Proceedings, 2001. XIII, 277 pages. 2001.

Vol. 2241: M. Jünger, D. Naddef (Eds.), Computational Combinatorial Optimization. IX, 305 pages. 2001.

Vol. 2242: C.A. Lee (Ed.), Grid Computing – GRID 2001. Proceedings, 2001. XII, 185 pages. 2001.

Vol. 2243: G. Bertrand, A. Imiya, R. Klette (Eds.), Digital and Image Geometry. VII, 455 pages. 2001.

Vol. 2244: D. Bjørner, M. Broy, A.V. Zamulin (Eds.), Perspectives of System Informatics. Proceedings, 2001. XIII, 548 pages. 2001.

Vol. 2245: R. Hariharan, M. Mukund, V. Vinay (Eds.), FST TCS 2001: Foundations of Software Technology and Theoretical Computer Science. Proceedings, 2001. XI, 347 pages. 2001.

Vol. 2246: R. Falcone, M. Singh, Y.-H. Tan (Eds.), Trust in Cyber-societies. VIII, 195 pages. 2001. (Subseries LNAI).

Vol. 2247: C. P. Rangan, C. Ding (Eds.), Progress in Cryptology – INDOCRYPT 2001. Proceedings, 2001. XIII, 351 pages. 2001.

Vol. 2248: C. Boyd (Ed.), Advances in Cryptology – ASIACRYPT 2001. Proceedings, 2001. XI, 603 pages. 2001.

Vol. 2249: K. Nagi, Transactional Agents. XVI, 205 pages. 2001.

Vol. 2250: R. Nieuwenhuis, A. Voronkov (Eds.), Logic for Programming, Artificial Intelligence, and Reasoning. Proceedings, 2001. XV, 738 pages. 2001. (Subseries LNAI).

Vol. 2251: Y.Y. Tang, V. Wickerhauser, P.C. Yuen, C.Li (Eds.), Wavelet Analysis and Its Applications. Proceedings, 2001. XIII, 450 pages. 2001.

Vol. 2252: J. Liu, P.C. Yuen, C. Li, J. Ng, T. Ishida (Eds.), Active Media Technology. Proceedings, 2001. XII, 402 pages. 2001.

Vol. 2253: T. Terano, T. Nishida, A. Namatame, S. Tsumoto, Y. Ohsawa, T. Washio (Eds.), New Frontiers in Artificial Intelligence. Proceedings, 2001. XXVII, 553 pages. 2001. (Subseries LNAI).

Vol. 2254: M.R. Little, L. Nigay (Eds.), Engineering for Human-Computer Interaction. Proceedings, 2001. XI, 359 pages. 2001.

Vol. 2256: M. Stumptner, D. Corbett, M. Brooks (Eds.), AI 2001: Advances in Artificial Intelligence. Proceedings, 2001. XII, 666 pages. 2001. (Subseries LNAI).

Vol. 2258: P. Brazdil, A. Jorge (Eds.), Progress in Artificial Intelligence. Proceedings, 2001. XII, 418 pages. 2001. (Subseries LNAI).

Vol. 2259: S. Vaudenay, A.M. Youssef (Eds.), Selected Areas in Cryptography. Proceedings, 2001. XI, 359 pages. 2001.

Vol. 2260: B. Honary (Ed.), Cryptography and Coding. Proceedings, 2001. IX, 416 pages. 2001.

Vol. 2264: K. Steinhöfel (Ed.), Stochastic Algorithms: Foundations and Applications. Proceedings, 2001. VIII, 203 pages. 2001.